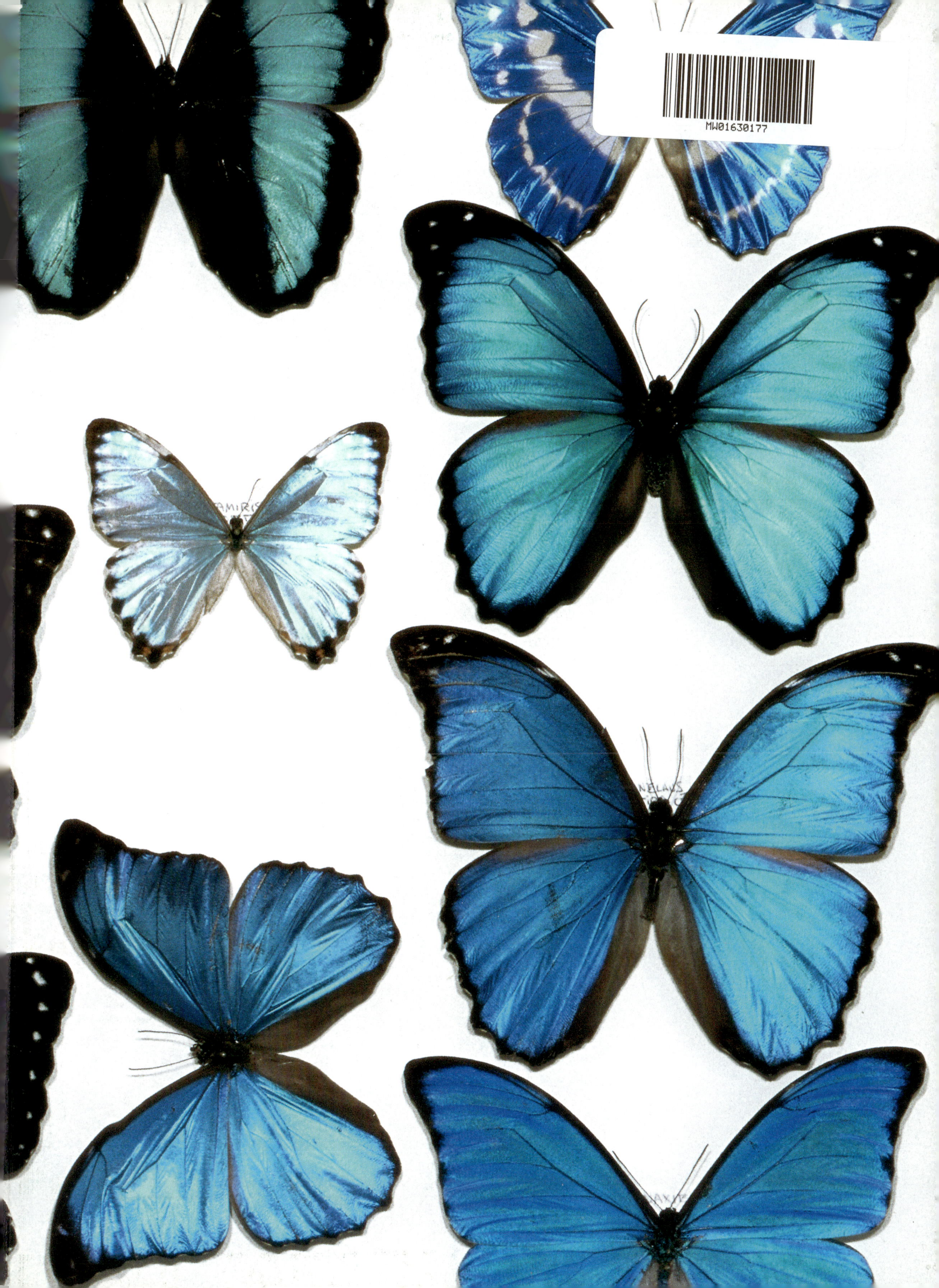
MW01630177

Printed in Canada

Library of Congress Cataloging-in-Publication Data

Tonnancour, Jacques de.
    [Insectes. English]
    Insects revealed : monsters or marvels? / Jacques de Tonnancour ; foreword by Sue Hubbell ; translated by Luke Sandford.
        p. cm.
    Includes bibliographical references (p.).
    ISBN 0-8014-4023-8 (cloth : alk. paper)
    1. Insects.   2. Insects—Pictorial works.   I. Title.
    QL463.T6513 2002
    595.7—dc21   2001006944

1   3   5   7   9   Cloth printing   10   8   6   4   2

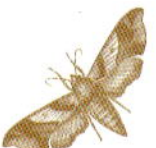

Preceding page: Luna moth *(Actias luna)* on a grass blade in the morning.

Insects Revealed

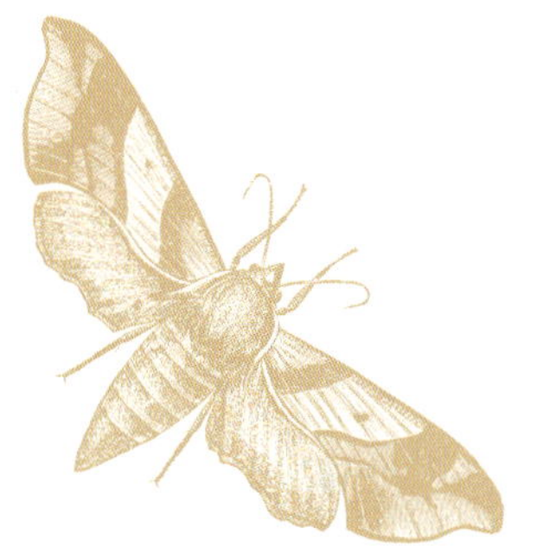

Monsters or Marvels?

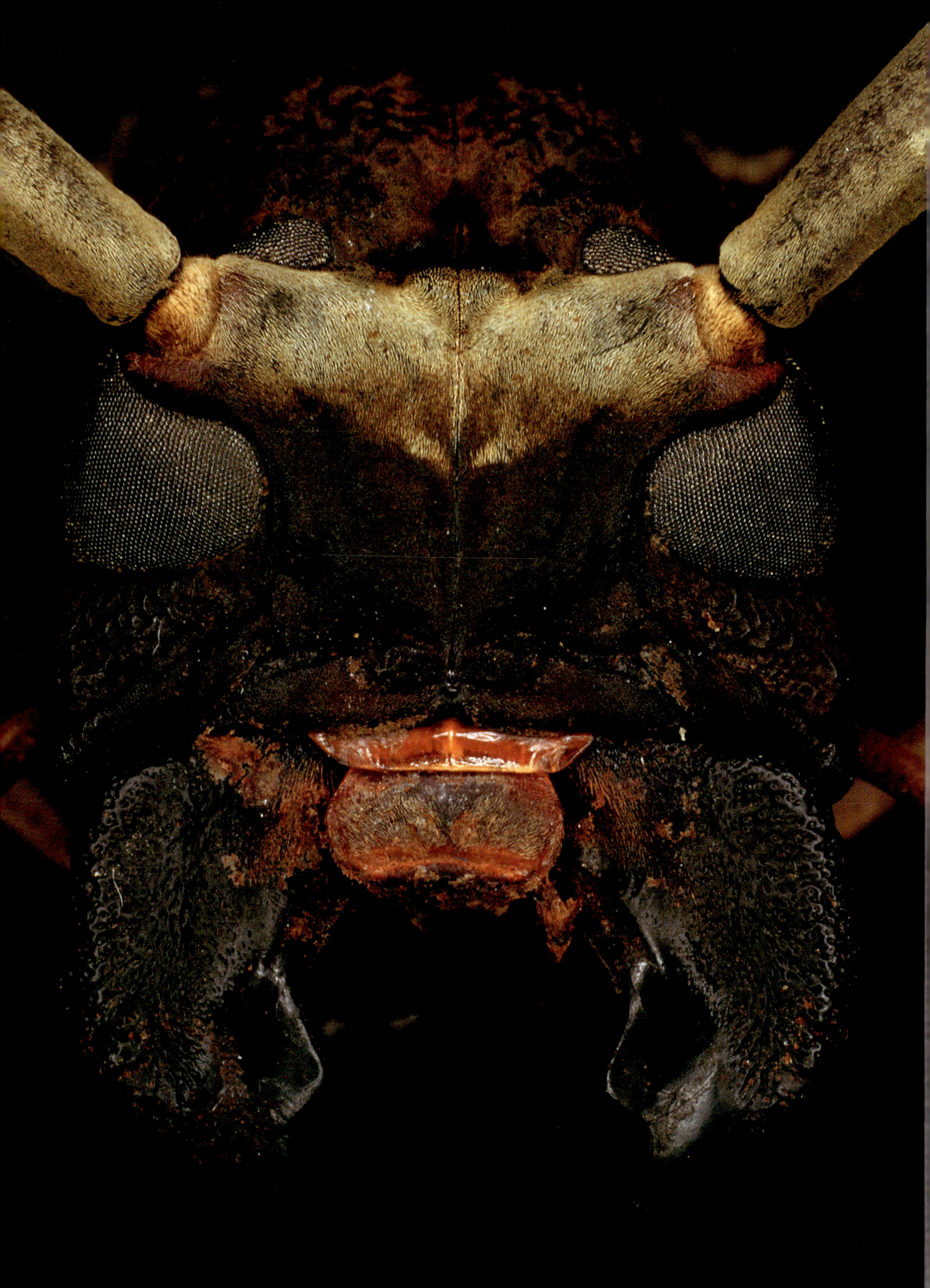

Jacques de Tonnancour

Insects Revealed

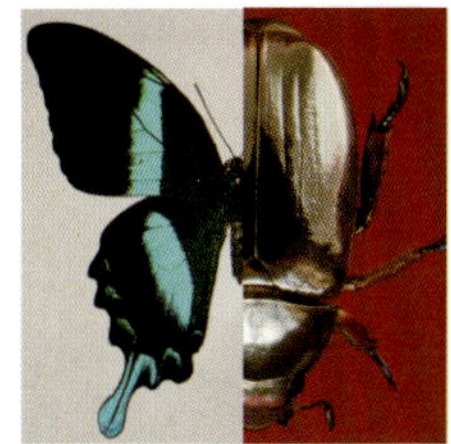

Monsters or Marvels?

Foreword by Sue Hubbell

Translated by Luke Sandford

Comstock Publishing Associates

a division of

Cornell University Press

Ithaca and London

Opposite page: A longhorn beetle (*Petrognatha gigas,* family Cerambycidae) from equatorial Africa.

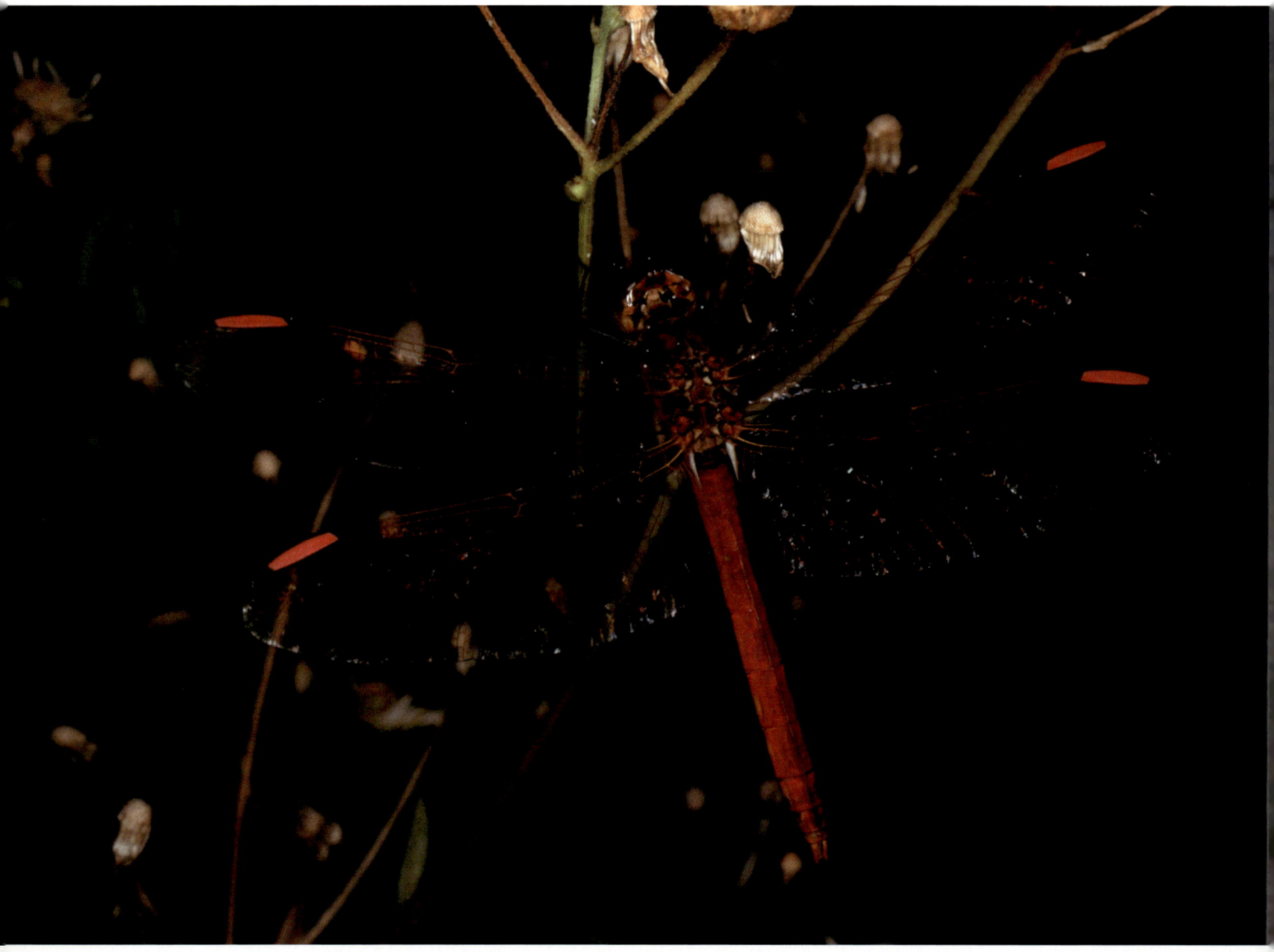

Dragonfly from Japan.

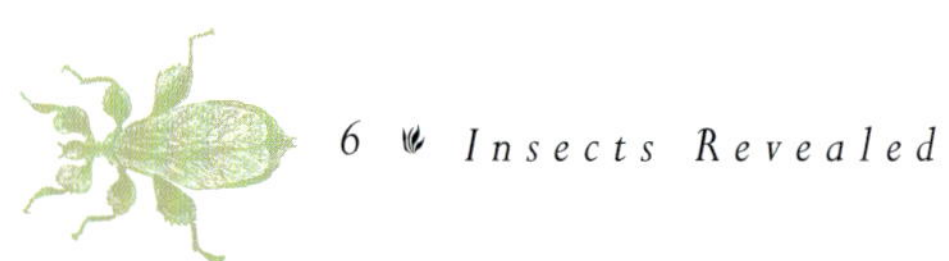

contents

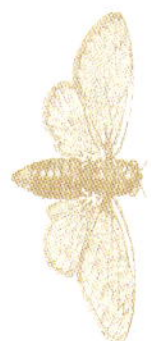

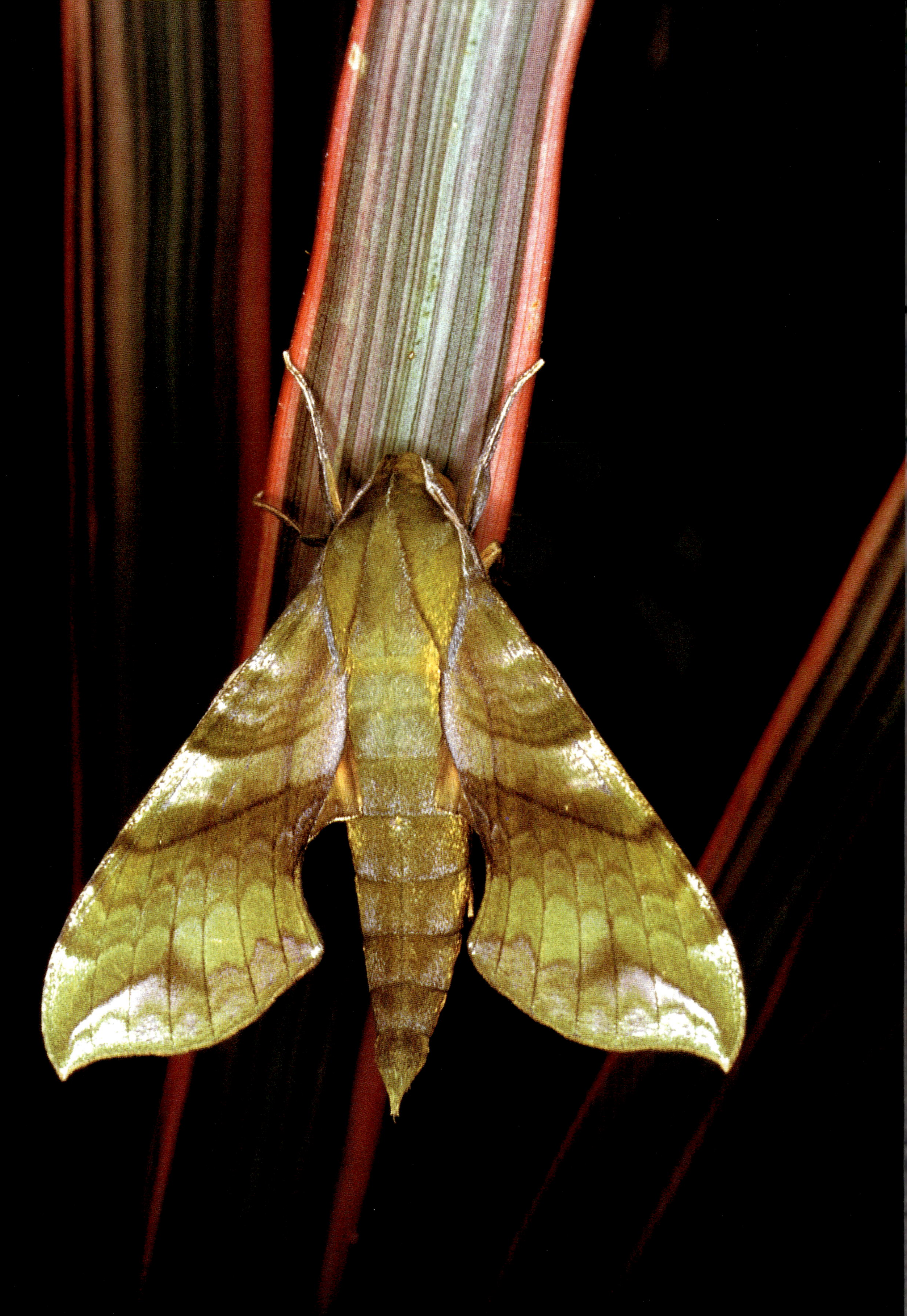

foreword

Years ago I wrote a magazine piece about three artists who used insects to make art, some of it beautiful, all of it provocative. In the course of writing I interviewed a number of those who had seen the artists' work in gallery shows. People responded in different ways, but all made the point that they now saw the world in a new way. It often takes an artist, such as Jacques de Tonnancour, to let us see beauty and to provoke our interest in the everyday parts of the world to which we are oblivious.

De Tonnancour, a respected artist from Quebec, put aside his paints twenty years ago and took up the camera in order to photograph insects and study their lives. He has, he confesses, a "lifelong fascination with insects." Those of us who share his fascination are always surprised by the fact that many people don't think of insects as everyday parts of our world. In fact they seldom even see them. The reason that this is surprising is that there are so many insects and they are all around us all the time.

Both in sheer numbers of individuals and in named and identified species, there are more insects than any other kind of animal. We have identified fewer than two million species of animals. Most of those are insects, and the majority of insects that live on the planet, entomologists tell us, have not been identified and named. Just beetles alone, something under four hundred thousand of them so far identified, represent 25 percent of the known animal kingdom, and one eminent beetle expert, Terry Erwin of the Smithsonian Institution, has estimated that there may be ten million species of beetles now living.

To be sure some insects are small and inconspicuous, but many are not. This past summer I noticed that here in the northeastern United States where I live there was a heartening increase in the lovely (and big) luna moths, one of which adorns the first page of this book. I was heartened because this part of the country was heavily sprayed with insecticides to kill gypsy moths in years past; those sprays killed many other lepidopterans in addition to the gypsy moths. The luna moths were particularly hard hit. I saw a number of them this summer, but when I asked neighbors if they, too, had seen them, they looked blank even when I described what they looked like. These are moths with wingspans of 7.5 to 10.5 cm. (3 to 4 in.) and their striking pale green beauty seems to me hard to miss when they alight on a screened window at night, attracted by light. I even found several in the road in the mornings, elegantly intact but dead, presumably victims of the night's oncoming automobiles. None of my morning walking companions noticed them until I pointed them out.

So we need the de Tonnancours of the world to bring little-noticed but common beauty to our attention. And perhaps, through beauty, to engage us more fully with the world around us.

Sue Hubbell

Opposite page: A pluto sphinx moth from tropical America (*Xylophanes pluto,* family Sphingidae).

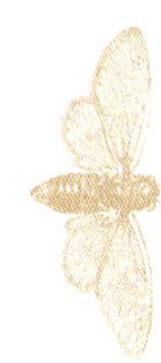

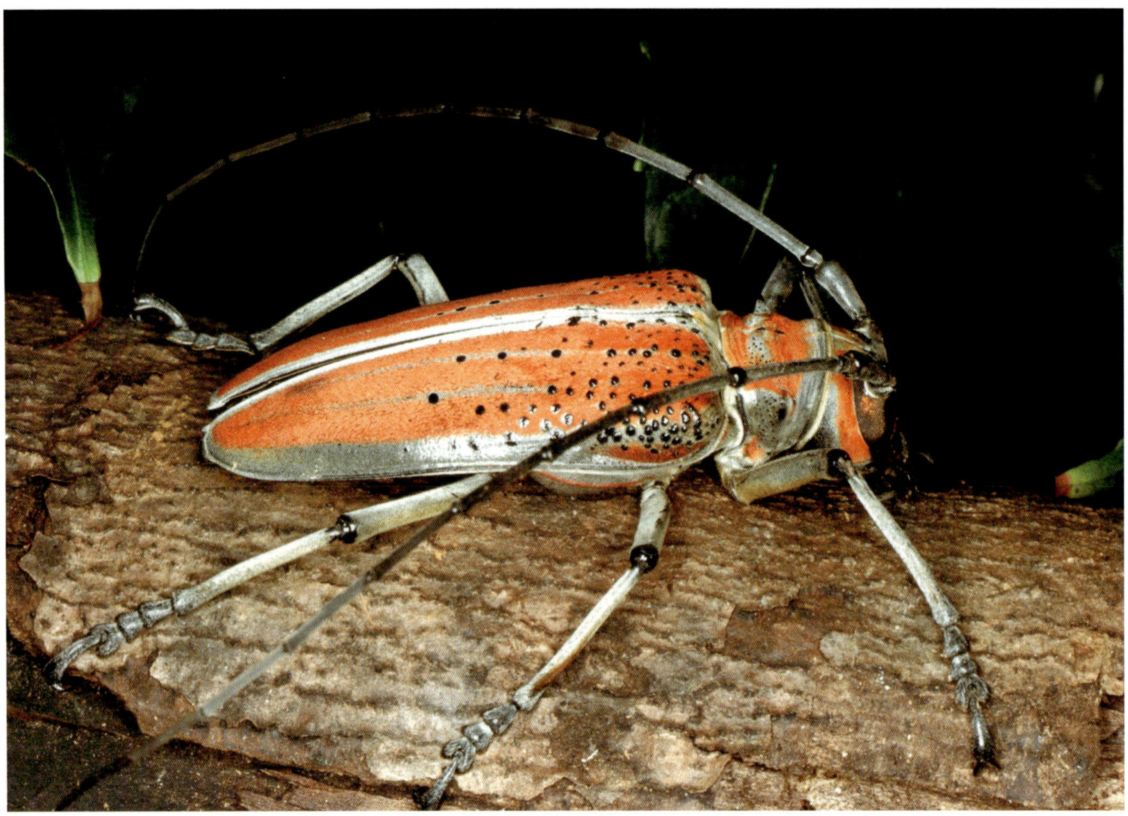

A longhorn beetle from Papua New Guinea (*Rosenbergia* sp., family Cerambycidae).

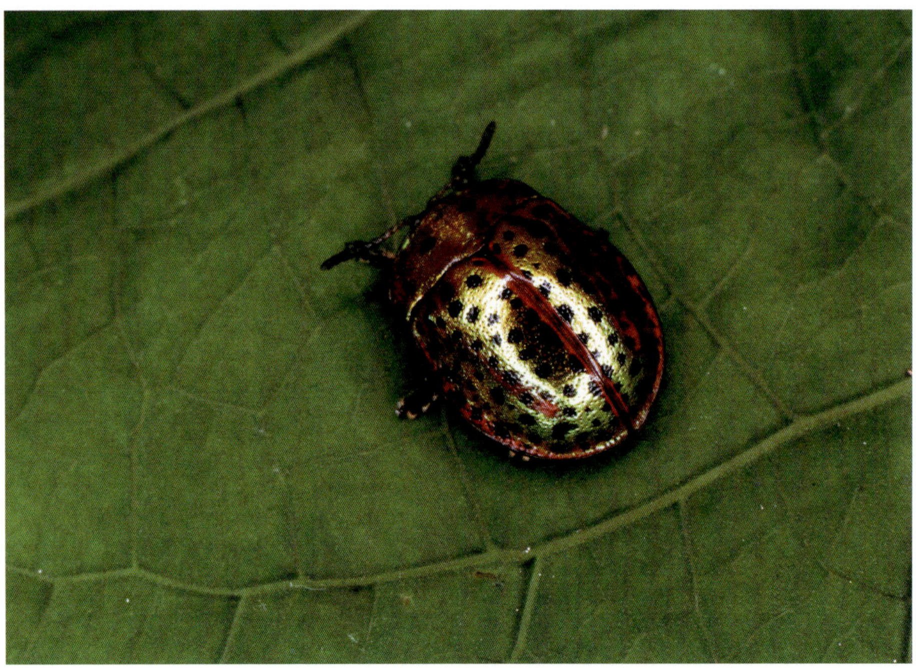

Chrysomelid beetle from the Dominican Republic. Children instinctively adore insects, especially those that look like toys.

preface

As one might expect, the tone and character of this book are shaped by my own personality and endeavors. I am a lover and collector of insects, a professional painter, and a photographer of nature, especially of insects. And if I were able to choose the readers of this book, they would ideally be people as yet unfamiliar with the world of insects, who might even find many insects unappealing, but who remain open-minded enough to be seduced by their mysterious beauty.

Children instinctively adore insects; I would like to reawaken this same sentiment in adult readers whose childhood memories are still within reach.

After browsing through these pages and viewing a few photographs, readers will realize that although this book offers a poetic or esthetic vision of the insect world, it also describes insects from a scientific point of view.

For too long entomologists have kept insects all to themselves. Their collections, housed in natural history museums, are their exclusive preserve. Indeed, if members of the public were granted access to present-day insect collections, these would hold as little interest for them as a twenty-volume encyclopedia. In order to foster public awareness of the extraordinary world of insects, one should take an anthological rather than an encyclopedic approach. Moreover, the presentation should serve to highlight the most attractive aspects of each type of insect.

Most books on entomology begin by placing insects within the taxonomic hierarchy and describing their morphology and physiology, along with countless other interesting facts. This traditional path, although appropriate for the subject, is not one that the uninitiated will necessarily enjoy.

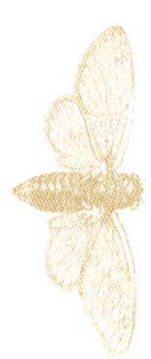

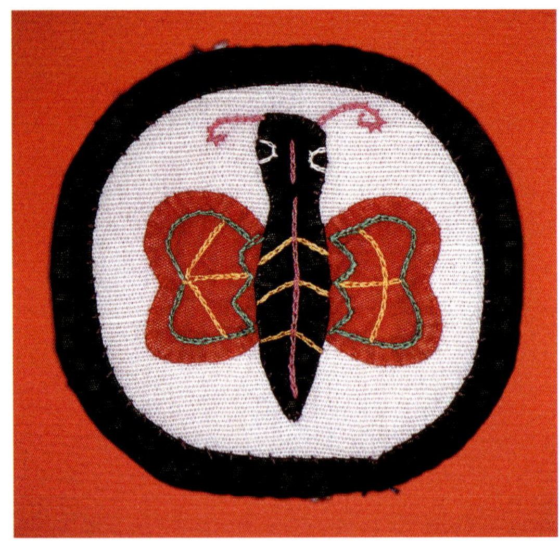

Mola embroidery made by the Cuna people of Panama.

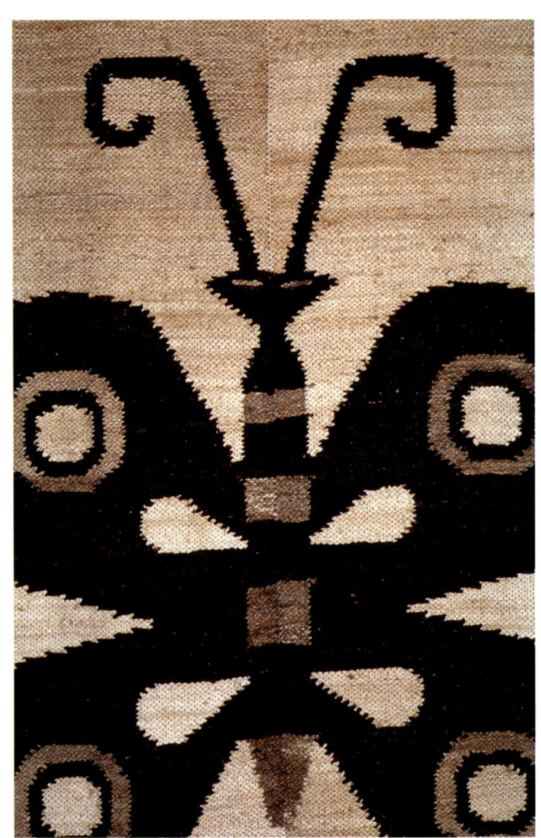

Wall tapestry from contemporary Peru.

In both these examples, the insect is more imaginary than real.

Before proceeding with a more objective treatment of the insect world, I would first like to explore the images that people have formed of insects across the ages. These representations can help us understand the image that we still have today of what insects are, or appear to be. In truth, these age-old depictions of insects have undergone little change; thus, in the public at large, insects are still misunderstood. If people are not yet familiar with the world of insects, what recourse do they have other than to give free rein to their imaginations? As a result, the insect created by the human imagination has long overshadowed the real insect.

If we suddenly transformed imaginary insects into real insects, perhaps readers would discover creatures even more extraordinary than those described in our ancient legends and folktales. Indeed, this book is founded on that very hope. Technical language is thus used sparingly in these pages so as not to discourage any well-intentioned readers, who must be drawn in before their education can begin.

All of the photographs in this book were taken by the author (unless otherwise indicated), and most of the insects represented here come from my own collection. A few of the specimens were generously provided by my colleagues; their cooperation is greatly appreciated.

I owe special thanks to Georges Brossard, Henri Miquet-Sage, Gontran Drouin, Gilles Deslisle, my son Pierre, the management and staff of the Montreal Insectarium, and to many others who provided me with unforeseen opportunities to photograph rare specimens that I might otherwise never have found.

My warm thanks also go to my old friend Jean-Pierre Bourassa, a professor at the University of Quebec at Trois-Rivières and a specialist in stinging insects, who eagerly offered to read the manuscript. Professor Bourassa's corrections and suggestions helped calm the misgivings that may afflict any nonspecialist who dares to tackle a topic as well established as that of the insect world.

I should mention that the descriptions that follow are in neither taxonomic nor geographic order. Each chapter deliberately brings together species from highly distinct orders and from families found in diverse and far-flung locales. The result is an anthology of some of the world's most beautiful, most peculiar, and most fascinating insects. The one hundred or so insects discussed in these pages, however, represent only a fraction of the total number—more than one million species worldwide.

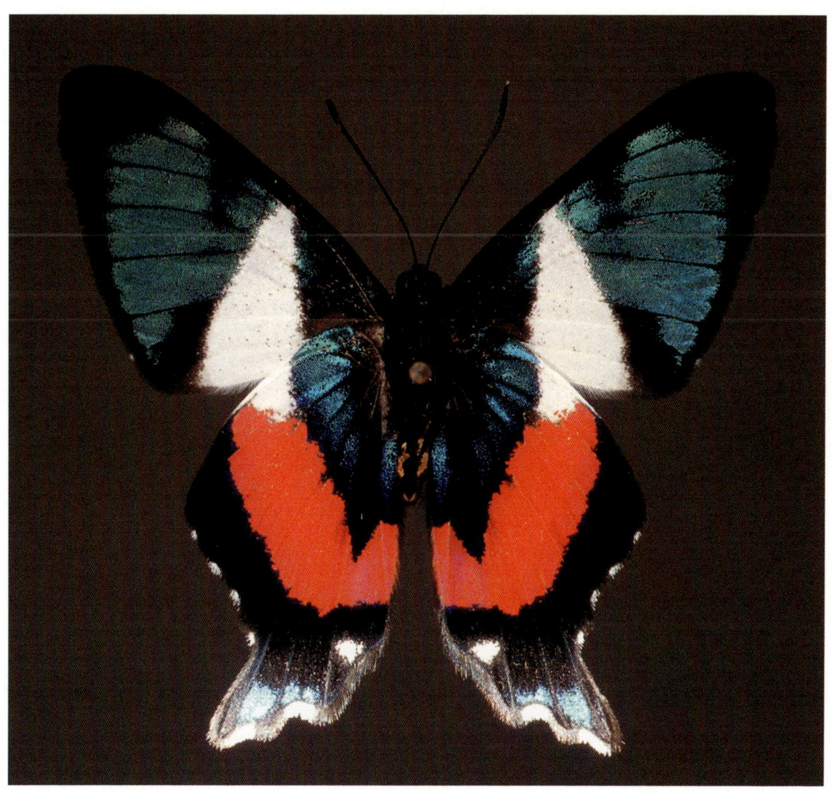

Metalmark butterfly (*Ancyluris formosissima*, family Riodinidae) from the Amazonian region of Peru. At play are two competing but nonetheless compatible conceptions of beauty: artistic and natural.

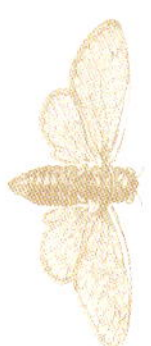

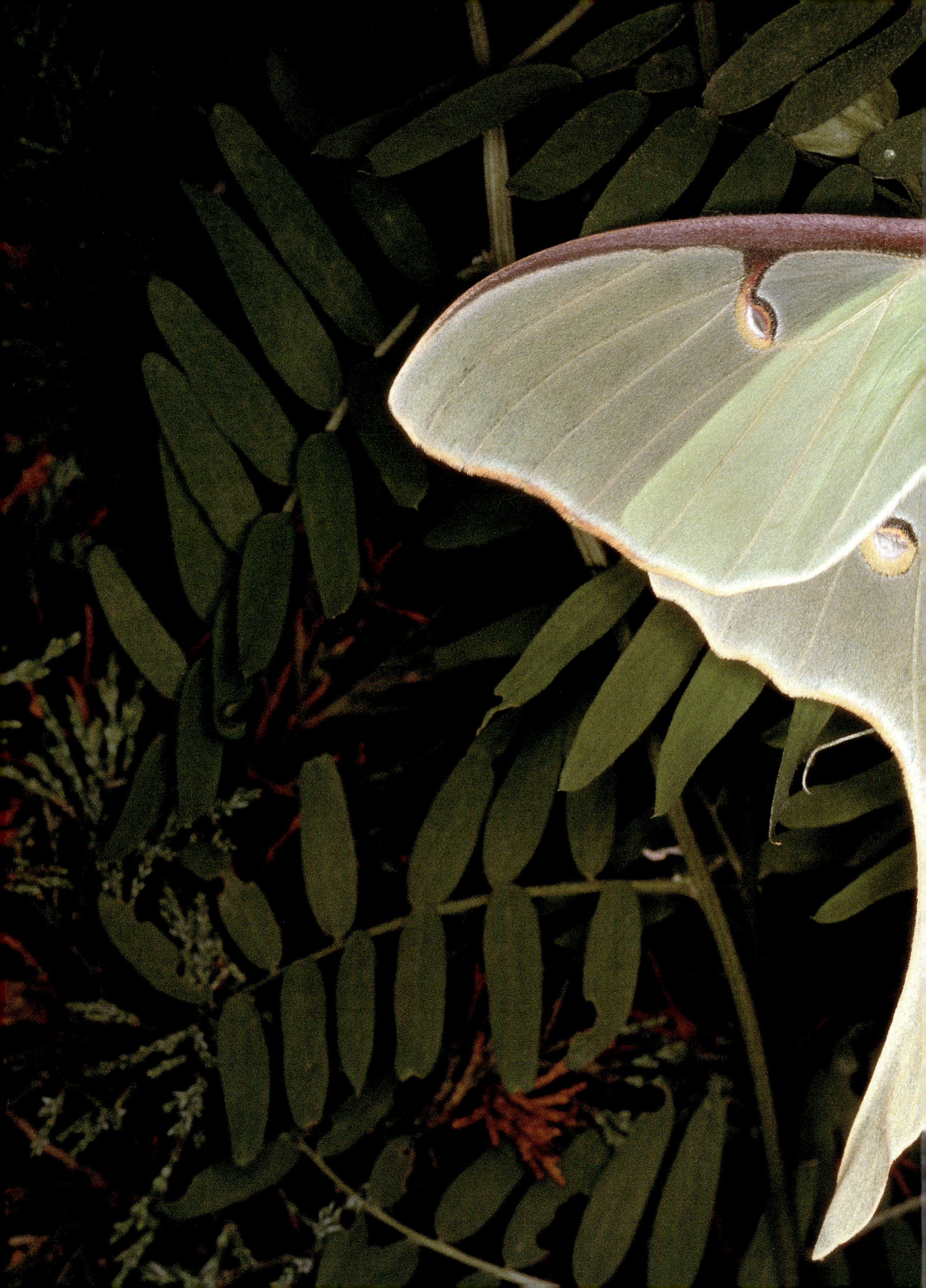

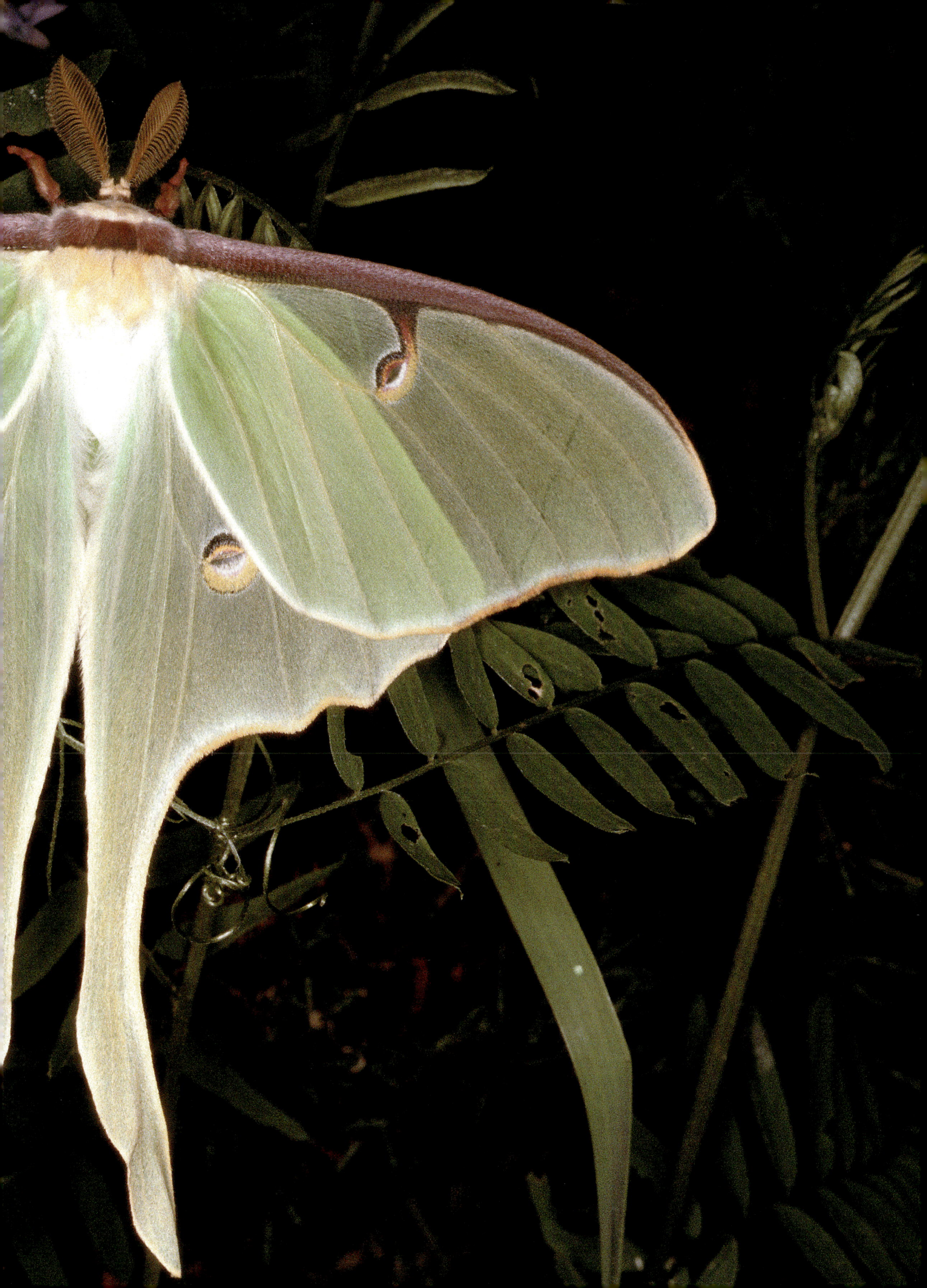

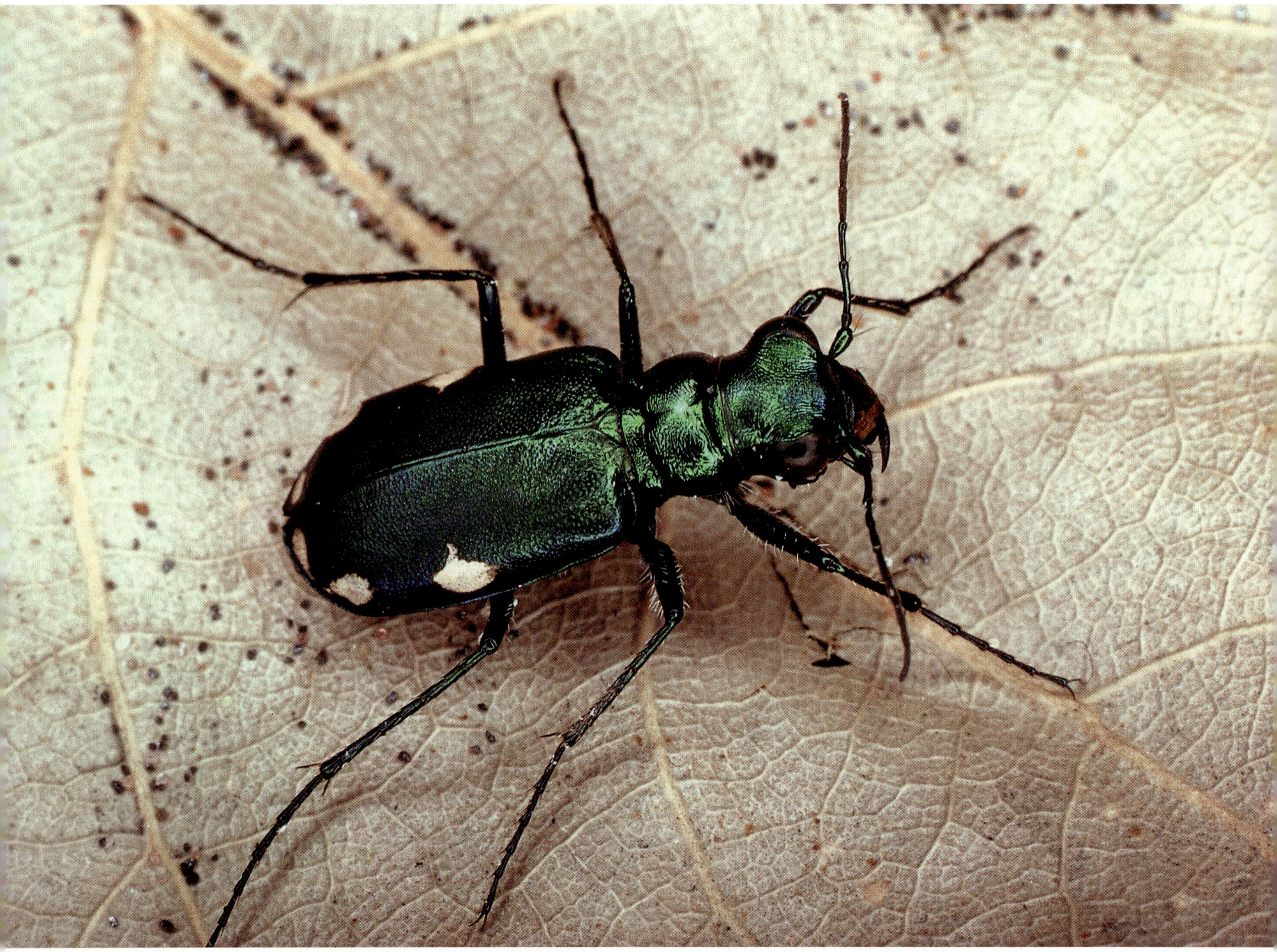

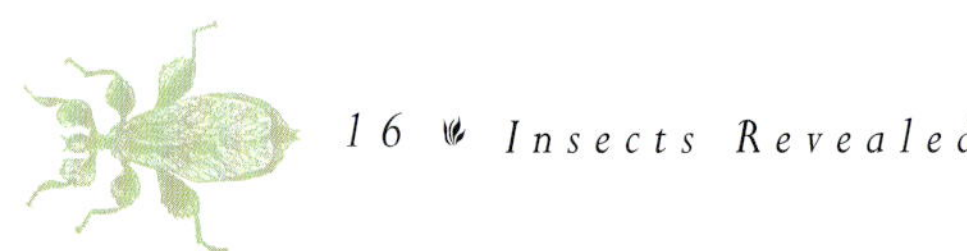

Above: *Cicindela sexguttata,* a superb species from the predatory tiger beetle family (Cicindelidae). This insect is found throughout eastern North America.

Preceding pages: Luna moth *(Actias luna),* a large member of the family Saturniidae. This is the only species of the *Actias* genus found outside Asia. Its range extends from southern Quebec to the southern United States, where it reproduces more than once per year. It is found in Quebec in late May and June. Wingspan: 9–10 cm. (3.5–4 in.).

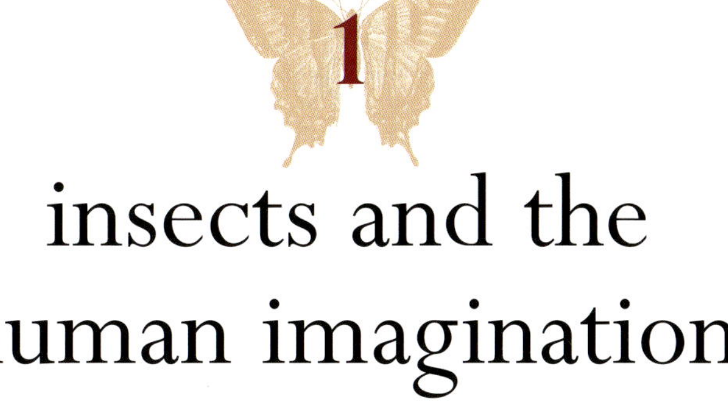

insects and the
human imagination

Owing to the urbanized lives we lead nowadays, our contacts with insects are becoming less and less frequent. Aside from those few species that coexist with humans, insects are no longer part of daily life for many of us.

This was not always the case. Human feelings toward insects have varied greatly over the ages and in different settings. To be sure, the response has often depended on the particular insect under consideration, but it has also been governed by differing cultural realities. Reactions have ranged from fear and loathing to affection and elation, even veneration.

Religious and popular beliefs, myths and legends have largely tended to assign insects to a world apart from our own—not all of them, of course, since most species do not affect us directly and thus go unnoticed. Those insects that do catch our eye for one reason or another often leave us amazed or awestruck by their strange, mysterious, and ghostly presence. A huge moth flutters beneath a streetlamp, an iridescent tiger beetle shimmers metallic blue and green in the middle of a rock-strewn trail, a hawk moth hums at twilight amid the lilac flowers—and just as suddenly it is gone.

Insects are both strange and strangers to us. Perhaps this is because their very make-up is so far removed from our own.

Insects are also mysterious because they know how to live among us while remaining invisible to our eyes. They cleverly take on the shape and color of various

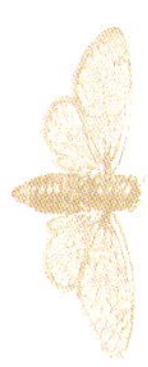

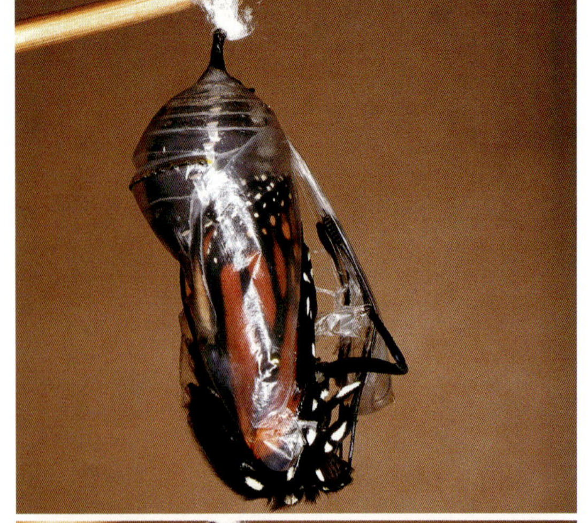

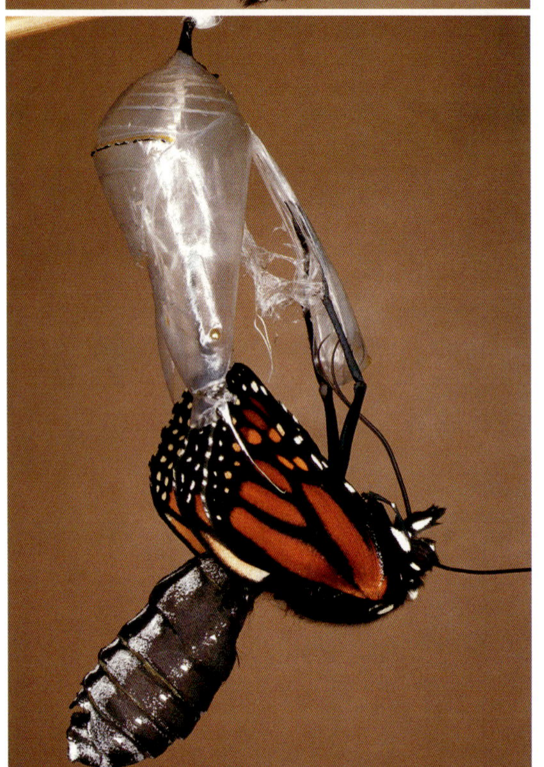

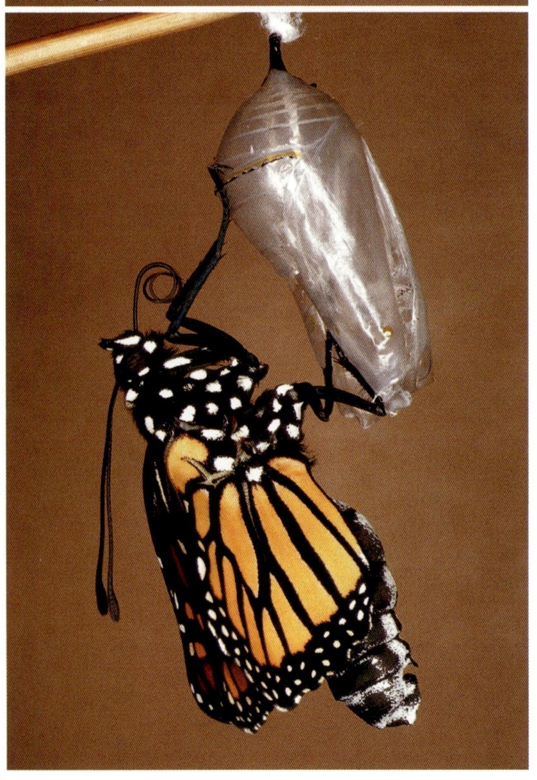

leaves, shriveled bark, or broken twigs. They appear and disappear in a trice, with a quick jump or a flash of their wings, emerging from nowhere, and quickly vanishing whence they came. As a result, many seem elusive and blessed with special powers, prompting us to wonder if they might be superior to us in some way.

Insects' mysterious qualities may also derive from their very specific power of metamorphosis. This phenomenon was more easily observable in the past, when more people lived in rural areas and were in intimate contact with a natural world that was less spoiled than it is today. Indeed, the ultimate magic trick in our most fantastic myths and legends is metamorphosis, or a complete change of form. Folklore from all parts of the world is full of gods, demons, fairies, and witches with the ability to transform us into frogs, deer, or ants.

Picture some children watching a butterfly emerge from a chrysalis: henceforth, the real world around them will contain the same fantasies and wonders that, until that moment, were the sole purview of legends and stories. But if the fantastic elements had not already existed in reality, how could they have become woven into our fairy tales?

It was in ancient times and primitive cultures that insects' magical properties gave rise to the most highly imaginative interpretations. These visions, although often very complex, nevertheless became quite disconnected from the external world.

The French poet and essayist Paul Valéry once remarked that even if a philosophical system is re-

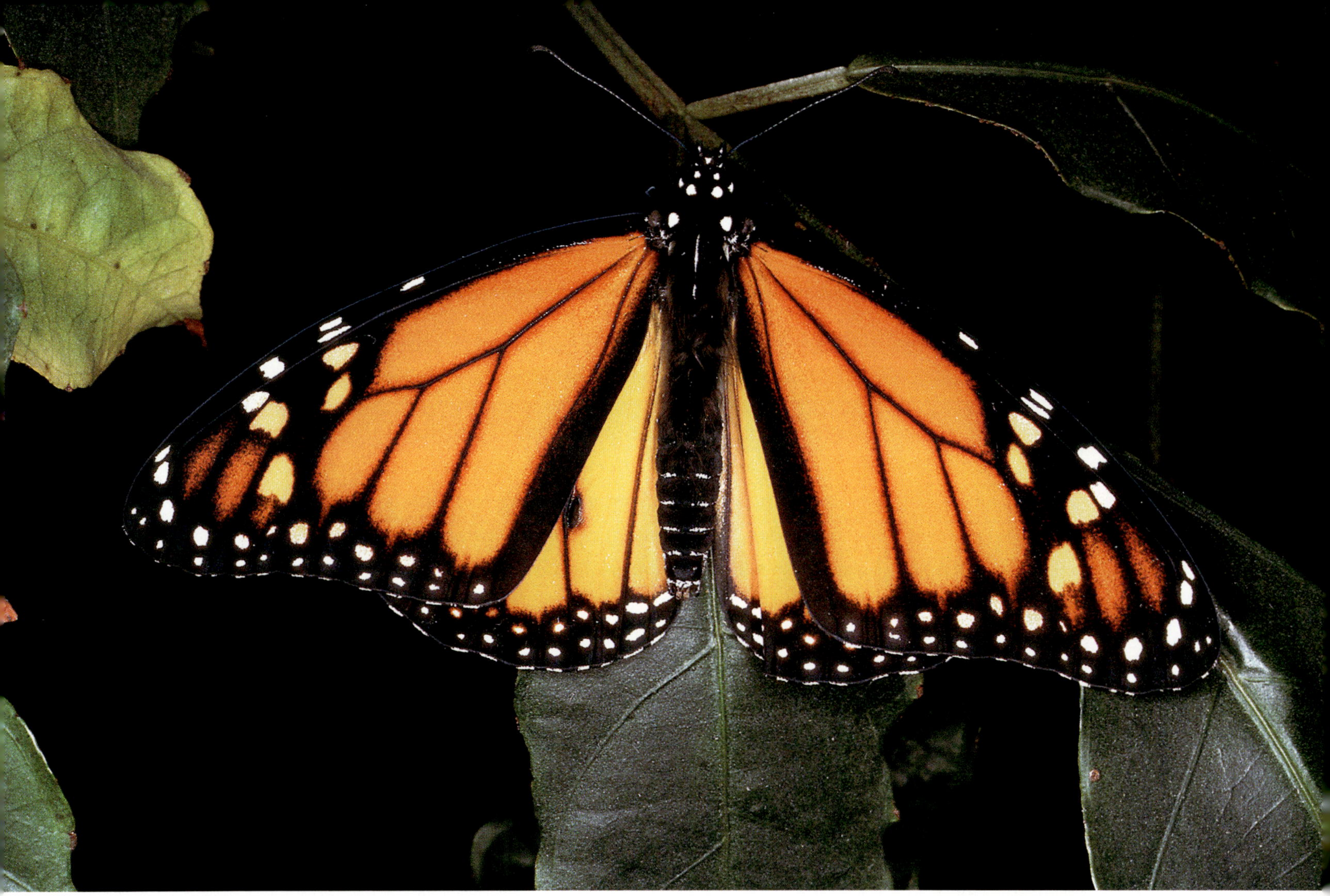

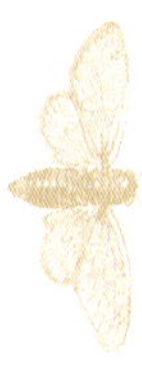

Above: The final stage of the metamorphosis of the monarch butterfly *(Danaus plexippus)*. This world-famous butterfly inhabits many regions with a range of climates.

Preceding page: The transparent shell of the chrysalis reveals the monarch butterfly's ultimate form and colors (top). During the two-week period when the chrysalis appeared to be dormant, not a second was wasted as the caterpillar miraculously transformed into a butterfly. The butterfly pushes off the chrysalis's covering and emerges virtually complete, stretching its legs and unfurling its proboscis (middle). It must still extend its tiny shriveled wings by pumping the hemolymphatic fluid through the wing's veins (bottom). The wings soon harden into a firm structure and the butterfly will then have completed the last phase of its development.

futed, it remains a work of art. The same applies to the fruits of "prescientific" thinking, which has bequeathed us wondrous stories and poetic philosophical systems. We have a heightened appreciation of them today since we do not have to accept them as true.

I have never quite understood my lifelong fascination with insects; as far back as I can remember, I have felt an irresistible attraction to them.

As a result, the behavior of people who react hysterically when they see or touch some tiny six-legged dot is simply unfathomable to me. Perhaps such different reactions depend on whether we experience a feeling of harmony or of conflict in our rela-

Metamorphosis

Considering all the nature programs available on television, you would think that everybody would know that insects pass through several stages before reaching their final adult form, the imago. Throughout metamorphosis, an insect takes on such different forms that it is hard to believe that they correspond to the same species; and indeed, there are still people who believe that caterpillars, chrysalises, and butterflies are distinct and unconnected creatures.

In the Dominican Republic, a gardener who often brings me insect specimens presented me with a caterpillar of the Papilionidae family. I thanked him, briefly mentioned the butterfly that would eventually emerge, and immediately felt the man's attention slipping away. Until then, he had always thought of me as sane, but the look on his face was unmistakable: it was simply inconceivable that caterpillars could become butterflies. For this man, a caterpillar was one type of insect, and a butterfly was another; it was as simple as that.

One day a woman speaking with the novelist and lepidopterist Vladimir Nabokov was describing a type of caterpillar found in her garden. Nabokov attempted to reassure the woman, saying that the caterpillar in question would become a swallowtail butterfly (family Papilionidae). "I don't think so," she replied, "I've never seen those caterpillars grow wings or anything else."

The French entomologist Rémy Chauvin relates this delightfully naïve interpretation of metamorphosis that he encountered during a visit to Corsica:

One day I was looking at some hazelnut trees that had been attacked by some type of caterpillar. An elderly shepherd came up to me and asked, "Are you looking at the hazelnut trees? They're quite sick this year and it's a terrible shame. Do you know how important hazelnut trees are in Corsica?"

"Of course I do," I replied. "Your hazelnut trees have been seriously damaged by caterpillars!"

"I think you're right," he replied. "But luckily in a few days a butterfly will show up that eats all the caterpillars. And then we'll be rid of them."

I stood there stunned. After thirty years [of research], some bizarre ideas still amaze me. I had to explain insect metamorphosis to my ill-informed companion and told him the story of the caterpillar that turns into a butterfly. He looked at me sadly as he put his grungy beret back on, saying: "You Frenchmen have always taken us Corsicans for fools."[1]

---

1. Rémy Chauvin, *Une étrange passion* (Paris: Le Pré aux Clercs, 1990).

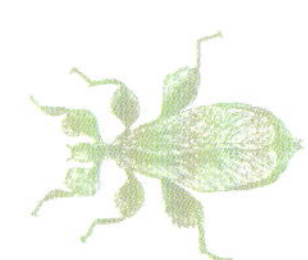

tionship with certain features of the world around us. In either case, which factor determines our reactions: genetic programming or cultural conditioning?

I also wonder why one culture would choose to revere a beetle, a butterfly, or some other insect, in effect adopting it as a vital symbol in its conception of the universe, while another culture from the same region may not ascribe any special status to it at all.

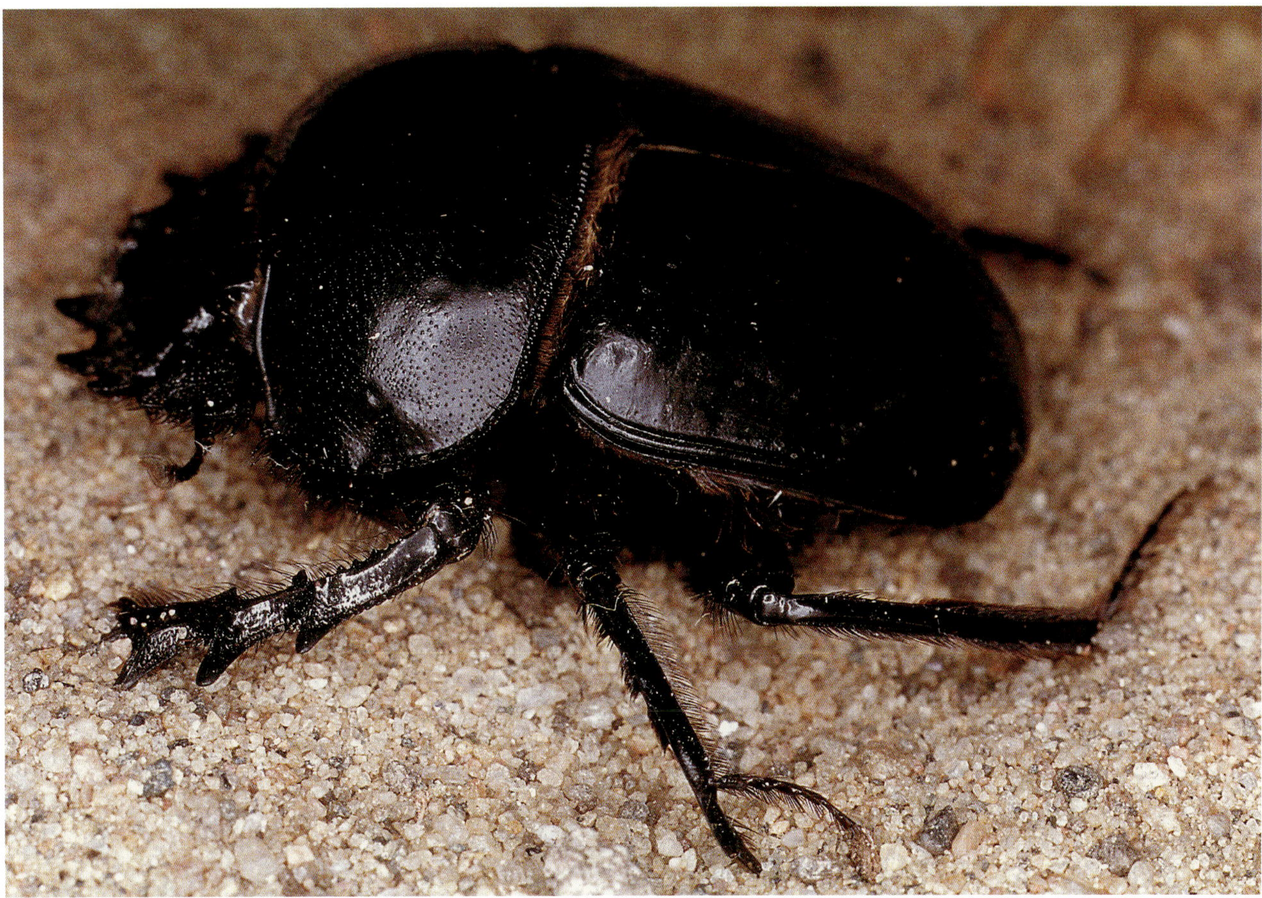

This beetle owes its illustrious name to Linnaeus, who named it *Scarabaeus sacer,* the sacred scarab beetle. These dung beetles provide a valuable service by burying animal droppings. They are found in the Mediterranean basin and in eastern Africa.

The ancient Egyptian veneration of the sacred scarab, or dung beetle (family Scarabaidae), was specific to that culture.[2] These insects enjoyed divine status, symbolizing Ra, the Sun god; as such, they embodied the notions of creation and destiny.

---

2. According to Bernhard Klausnitzer, Egyptians from different historical periods honored different species of scarabs, belonging to five distinct genera. The sacred scarab of the past was thus a composite insect; modern science, however, refers to it by the name that Linnaeus assigned to one species, *Scarabaeus sacer* (*Beetles* [New York: Exeter Books, Simon and Schuster, 1983]).

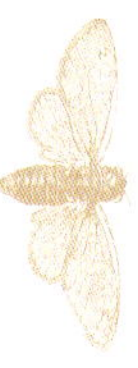

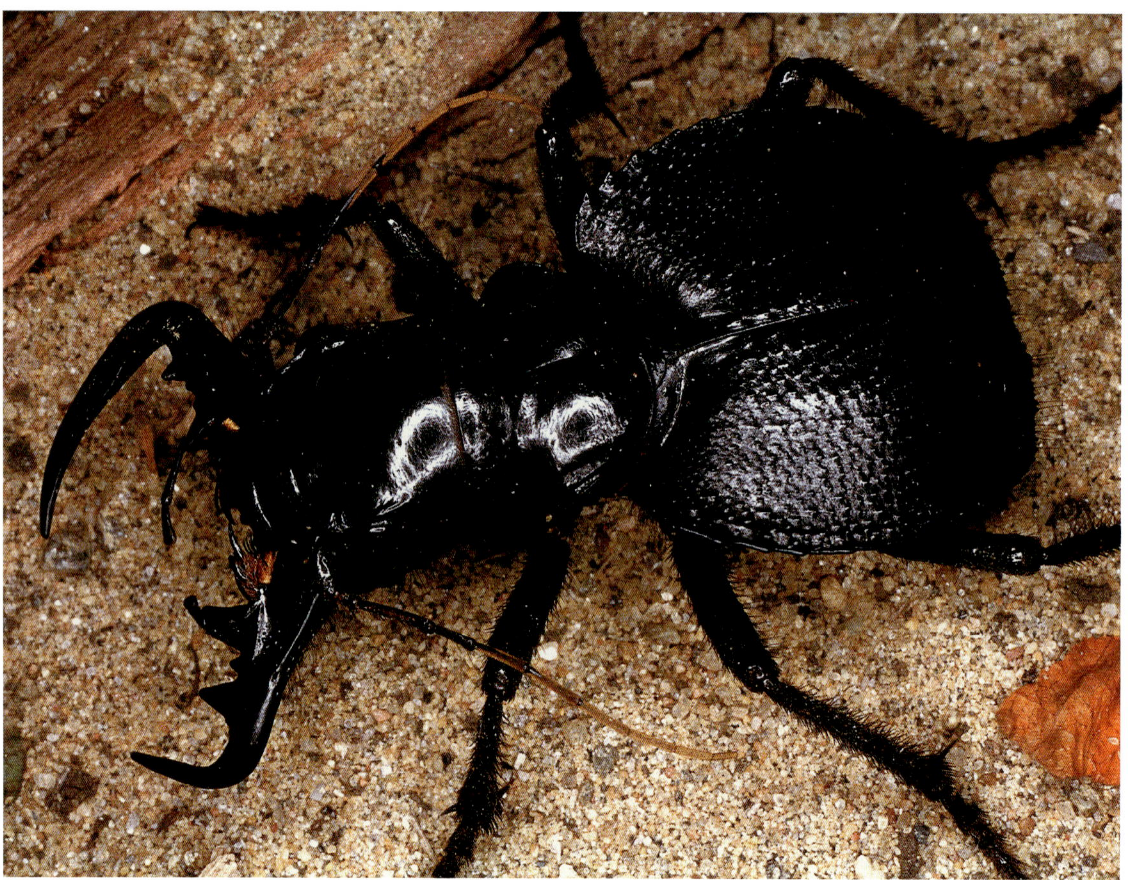

Some insects appear to vindicate Saint Augustine, who claimed that all insects are creatures of the devil.
Above: A tiger beetle in the genus *Manticora,* the largest members of the tiger beetle family. This ferocious wingless predator is indigenous to South Africa.

The ancient Egyptians attributed noble aspirations to these creatures, whose terrestrial function is as mundane as can be.

In other places where dung beetles were abundant, cultures responded differently to their presence and behavior. Unlike their Egyptian counterparts, the ancient Greeks treated dung beetles with contempt, considering them inimical to knowledge, art, and culture. In the Middle Ages, Christians' negative reactions hinged on questions of faith; for them, the dung beetle was symbolic of the sinner and of the heretic rolling his ball of unorthodox beliefs around the world. I remember reading once that Saint Augustine considered all insects to be creatures of the devil!

By way of contrast, insects that possess irresistible and poetic charms, such as butterflies, have won universal acclaim and have evoked virtually identical associations of images and ideas in different cultures throughout human history. These symbols are so strikingly similar that one could call them expressions of our col-

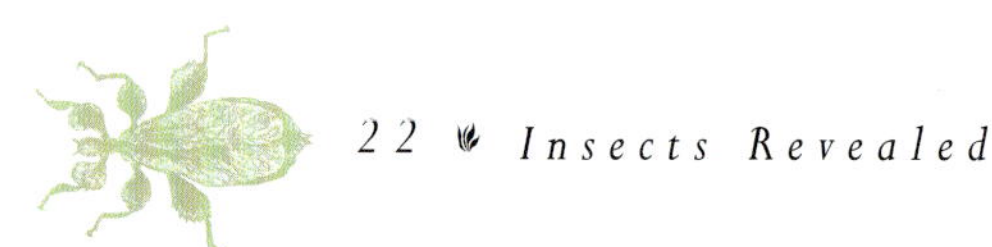

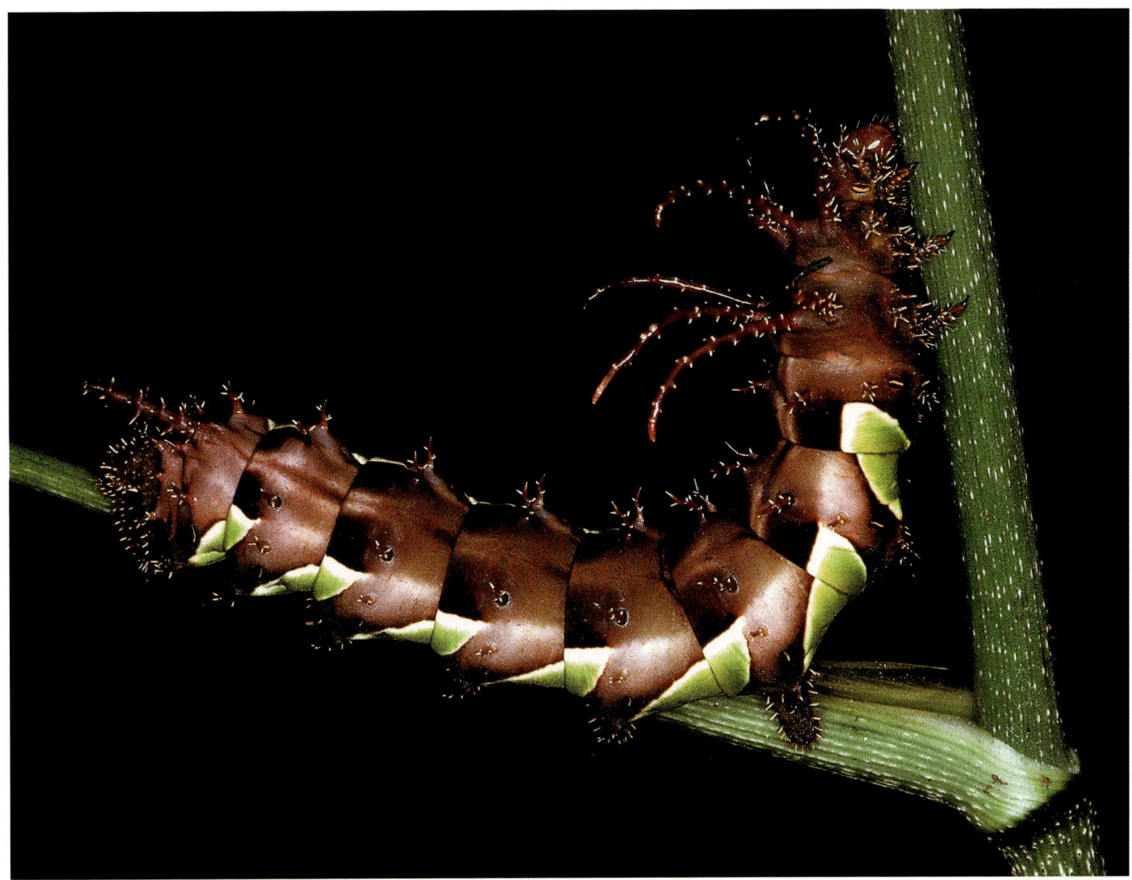

This saturniid caterpillar *(Citheronia hamifera)* is in the subfamily Ceratocampinae, found only in North and South America. Despite its malevolent appearance, the caterpillar is harmless. This species makes its home in tropical America.

lective unconscious. In this respect, I think in particular of the association between human souls and butterflies, which seems to be the most widespread notion.

The ancient Greeks believed that butterflies were the visible forms taken on by the souls of the departed during their journey to the afterlife. Given its simplicity, was this poetic image intended for the ears of children? Or perhaps some deeper connection was at work: several centuries later, Aristotle went on to combine this association within the single word *psyche,* signifying both "soul" and "butterfly."

Closer to our own times, we have the dramatic conclusion of the Swedish film *Elvira Madigan* (1966). Two lovers frolic against a background of wildflowers to the heavenly strains of Mozart. The couple has agreed to a suicide pact, but the scene is so luminous and captivating in its beauty that we forget what awaits them. As Elvira skips about a meadow, she catches a tiny white butterfly; just when she is about to let

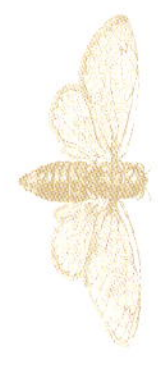

it go, a shot rings out, the music stops, and the camera lingers on her open hands as the butterfly flits away. Then a second gunshot shatters the endless silence, and we die along with the lovers. Never has a tiny white butterfly been invested with so much meaning.

Such symbols are not found only in Greece or Sweden, however; I believe that they belong to all times and to all places since they dwell within us all.

In the Amazon, caligo butterflies—a giant species active at twilight that has "owl eyes" on the underside of its posterior wings—often disturb those who watch them surveying the darkened doorway of some hut before entering and re-emerging to continue their inspections elsewhere. Some believe that the caligos are actually souls in search of their former homes.

My own observations of butterflies in the enchanting regions of the Amazonian forest suggest that this thought could occur naturally to anyone, and would not seem bizarre or naïve in any way.

Far from the Amazon, on the island of Madagascar, noctuid moths (family Noctuidae), which also frequently enter people's homes, have generated the same symbolic connection of moths and butterflies with human souls. According to the Madagascan entomologist Paul Griveaud, "The Madagascan legends relate that the 'lolopaty' represent the spirits of the deceased who have returned to visit the living. In fact, *lolo* in the Malagasy language means both spirits and butterflies."

It seems doubtful that these beliefs could have traveled all the way from ancient Greece to the Amazon or Madagascar. Perhaps they developed from the collective unconscious wherever the same underlying cause aroused them.

Across the ages, prejudices against moths have been commonplace. For many years, popular beliefs assigned noble and sympathetic roles to butterflies (diurnal Lepidoptera) and generally sinister and hostile roles to moths (nocturnal Lepidoptera). The Roman poet Ovid, in his *Metamorphoses,* described moths as "funereal." This is tantamount to a value judgment that grouped moths together with bats, owls, and other creatures considered bad omens. Even today, if a butterfly enters a

Opposite page:
The owl butterfly (genus Caligo, family Brassolidae) is indigenous to tropical America.

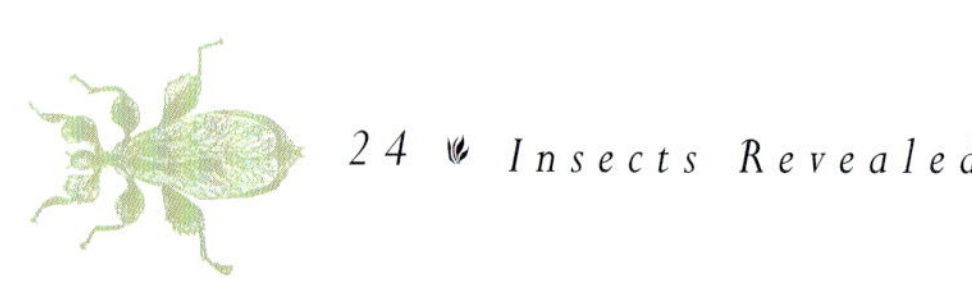

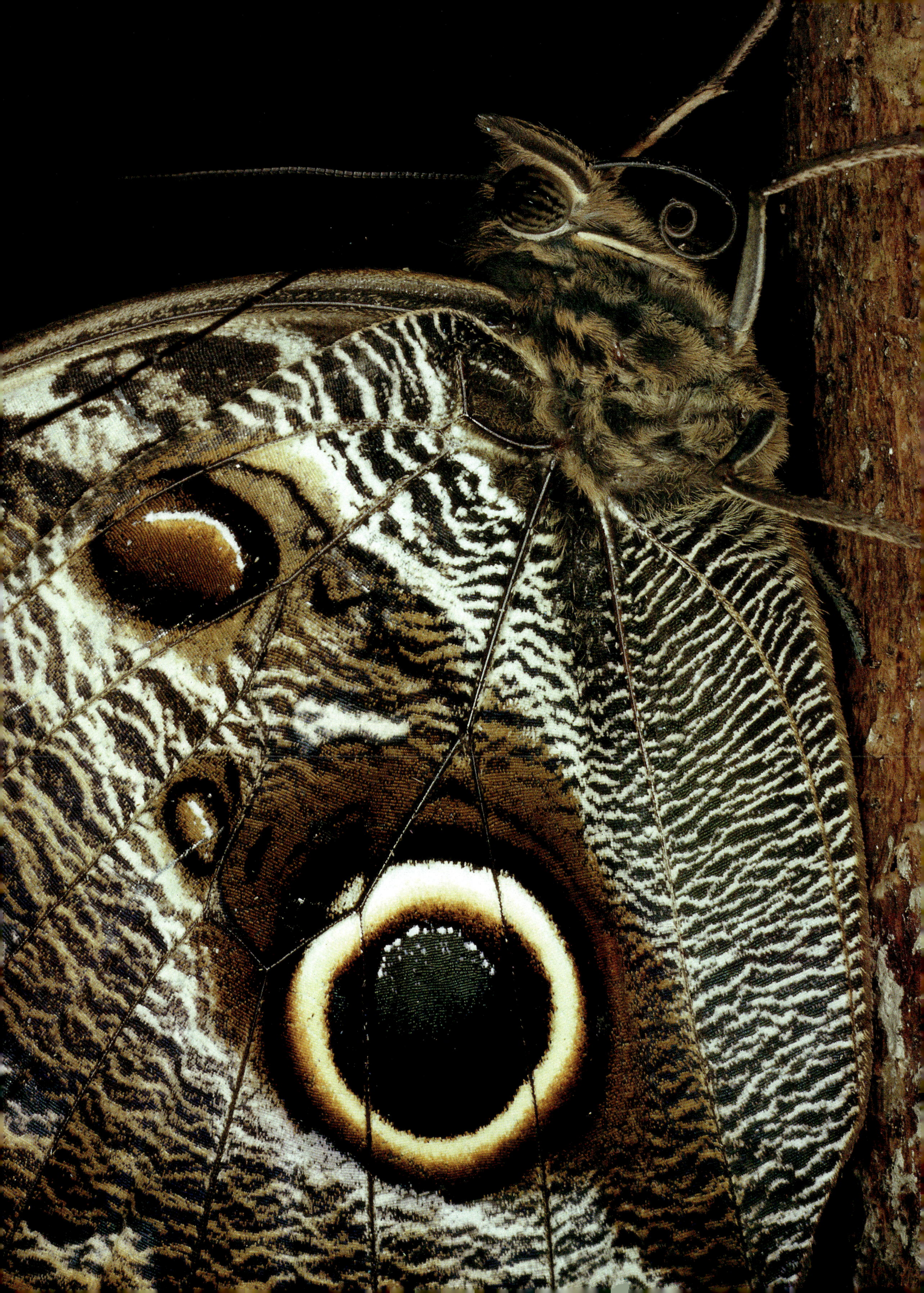

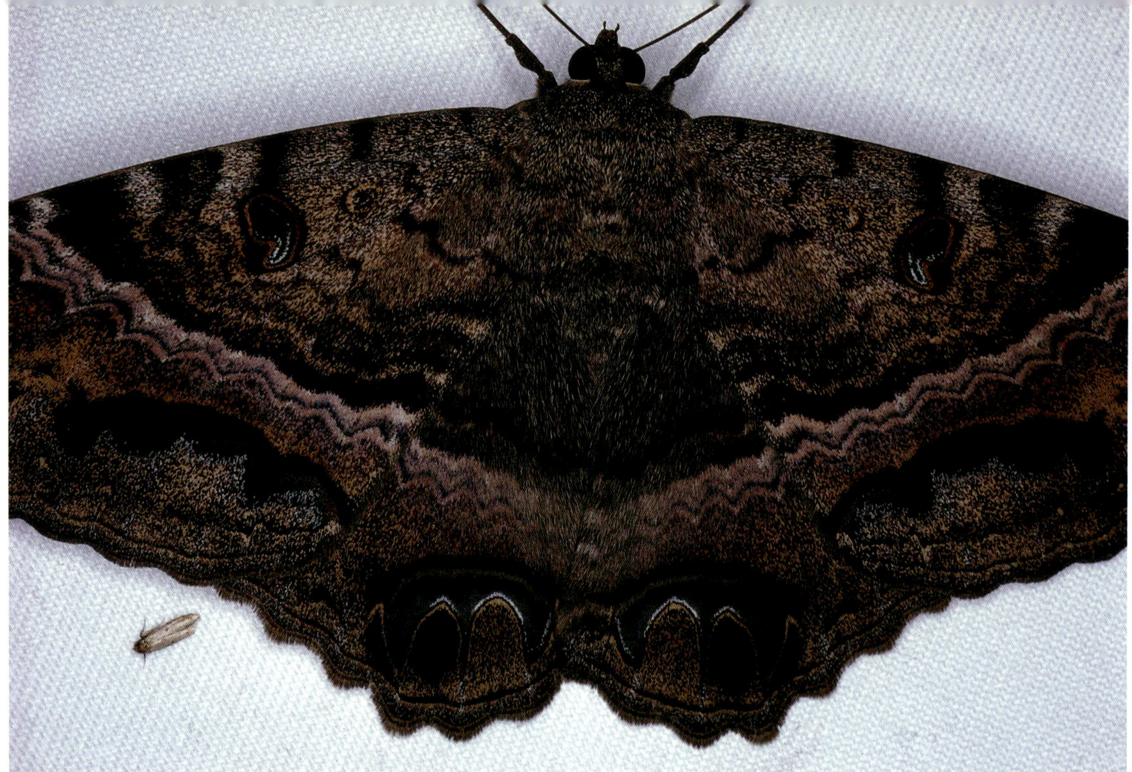

home, everyone is delighted. But if an impressively large moth does the same thing, people begin to fear a death in the family. This reaction is unwarranted and happens only because moths are usually darker colored than butterflies. Their coloration helps moths to escape the clutches of the diurnal predators that try to flush them out while they sleep, well camouflaged on tree trunks or in leaf litter. Neither Ovid nor popular opinion ever stopped to consider this aspect of moths' finery.

The enthusiast, however, finds extravagant beauty in moths, a group of insects that most people never catch sight of. Moreover, there are a great many varieties; in fact, they are ten times more numerous than butterflies. When I examine my light trap, I often feel very alone as I marvel at these inexhaustible wonders, which remain unknown to so many.

Ascalapha odorata, a large noctuid moth indigenous to the neotropics (the region of the New World extending southward from the Tropic of Cancer), provides us with an excellent example of a mistrusted species. Its nicknames leave no doubt as to its popular (or unpopular) standing: the "devil's butterfly," the "black witch," and many others, all very uncomplimentary. Why is this so? Perhaps because this moth is large and dark in color; it also flies haphazardly, and its eyes, which reflect even small amounts of light, seem illuminated from within, producing a reddish glow. Taken together, these traits are enough to turn a moth into a devil or a witch that it would be unwise to let into one's home. And what could the "death's-head moth" *(Acherontia lachesis)* foreshadow other than what its name indicates?

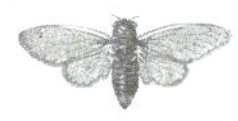

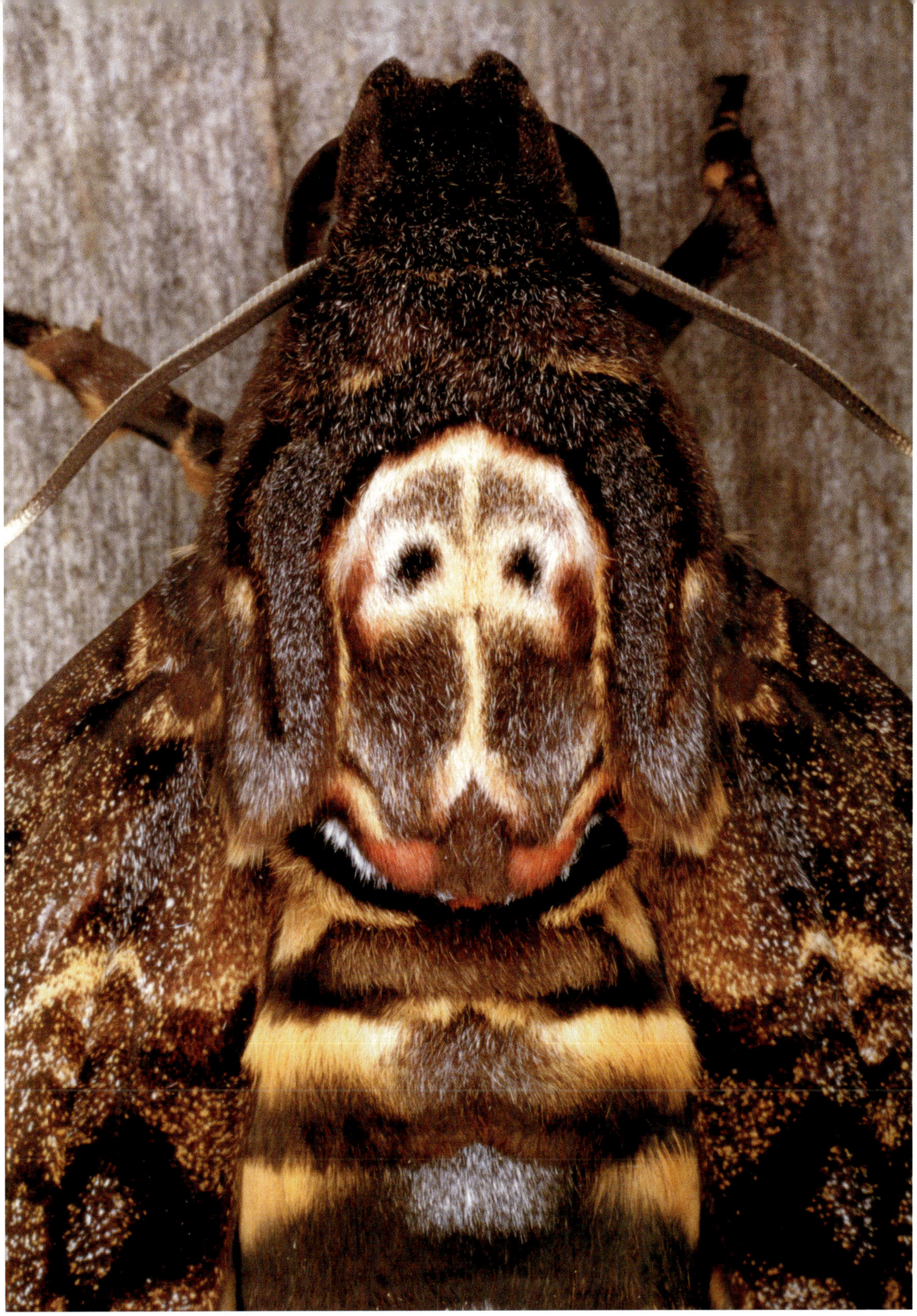

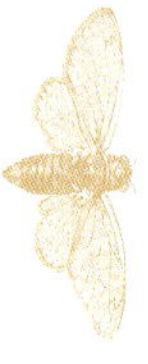

Some insects have a surrealistic aura that earns them mythic status. With their larger-than-life qualities, they are often ascribed magical powers. Two fine examples are these nocturnal Lepidoptera, *Ascalapha odorata* (opposite page), a large noctuid moth from the neotropics, and *Acherontia lachesis* (above), a sphingid from Southeast Asia. Their respective nicknames, "black witch" or "devil's butterfly," and "death's-head moth," bear witness to the negative impressions they leave.

Ascalapha odorata is one of the largest noctuids, measuring some 15 cm. (6 in.). Covered with dense hairs and lavishly ornamented, it is widespread throughout tropical America.

The genus *Acherontia* contains only three species, all of which feature skull-shaped patterns on the thorax. The name of the species shown here derives from Lachesis, one of the three Fates in Greek mythology, whose job was to measure the thread of human life so that it could be cut at the appropriate moment.

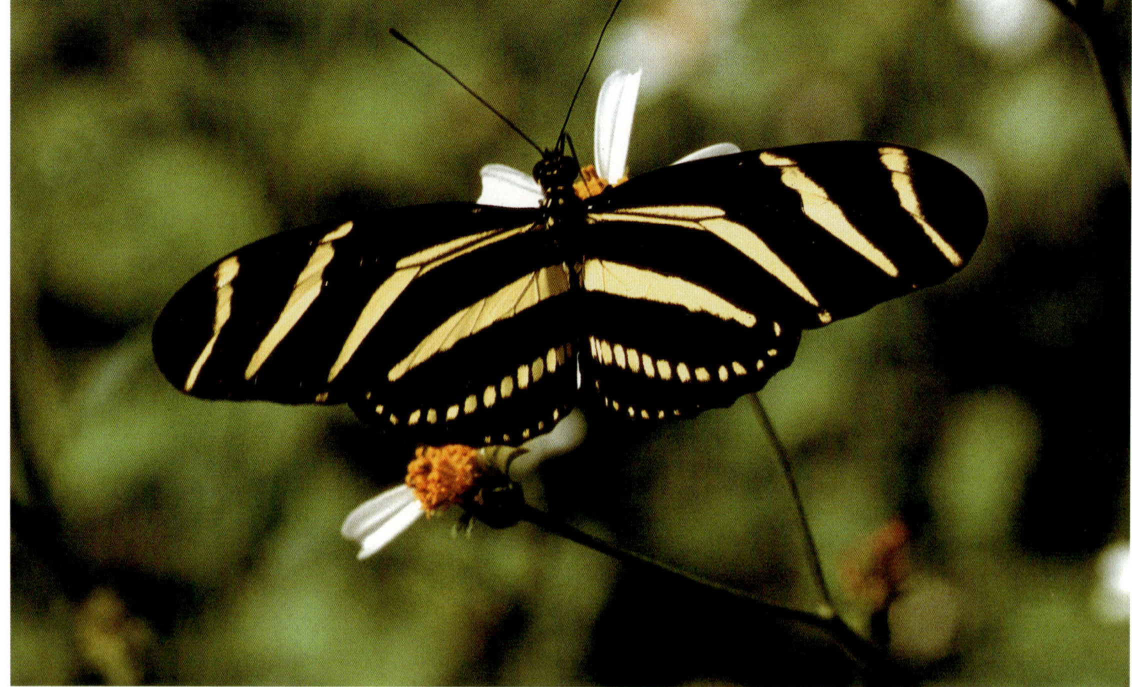

Humankind has always cherished butterflies as the most beautiful and most graceful objects in creation.
Above: Zebra longwing butterfly (*Heliconius charitonius,* family Heliconiidae). The range of this species extends from Florida to the northern regions of tropical America.

Humankind has historically been aloof and distant—even hostile—toward most insects. Nevertheless, some insects have been granted privileged status, owing to their beauty or strangeness. (Isn't beauty itself strange at times?) At other times, humans have recognized insects' obvious utility; more rarely, moral qualities have been ascribed to various insects, as in the case of the industrious ant.

Butterflies have always been viewed as nature's most beautiful and graceful creatures. Their apparent harmlessness has also served to enhance their standing in our eyes, a standing that other insects have not been able to attain. In the public's mind, butterflies are thus somehow distinct from other insects. People often ask me: "Other than butterflies, do you also collect insects?"

Insects' strangeness takes many forms. Luminescent insects are a class apart from the others, but only when they glow at night; during the day, they resume a more humdrum existence. Their nocturnal activities automatically link them with the hereafter; Quechuan speakers in the Andes call them *añañahui,* or "ghost eyes."

For millennia, humans have appreciated insects that are useful to them, whether as specialized producers, such as bees or silkworms, or in auxiliary roles, such as aphid-devouring ladybugs. Across all languages and cultures, ladybugs have been treated for centuries with more kindness than any other tiny creature. In the Nordic countries, the ladybug was dedicated to Freya, the goddess of love, whereas in some Christian countries it paid tribute to the Virgin Mary—hence the name "ladybug" in English, *bête à bon Dieu* ("God's creature") or *bête de la Vierge* ("Virgin-creature") in French, and *mariquita* ("little Mary") in Spanish. The cult of the ladybug may derive less from its

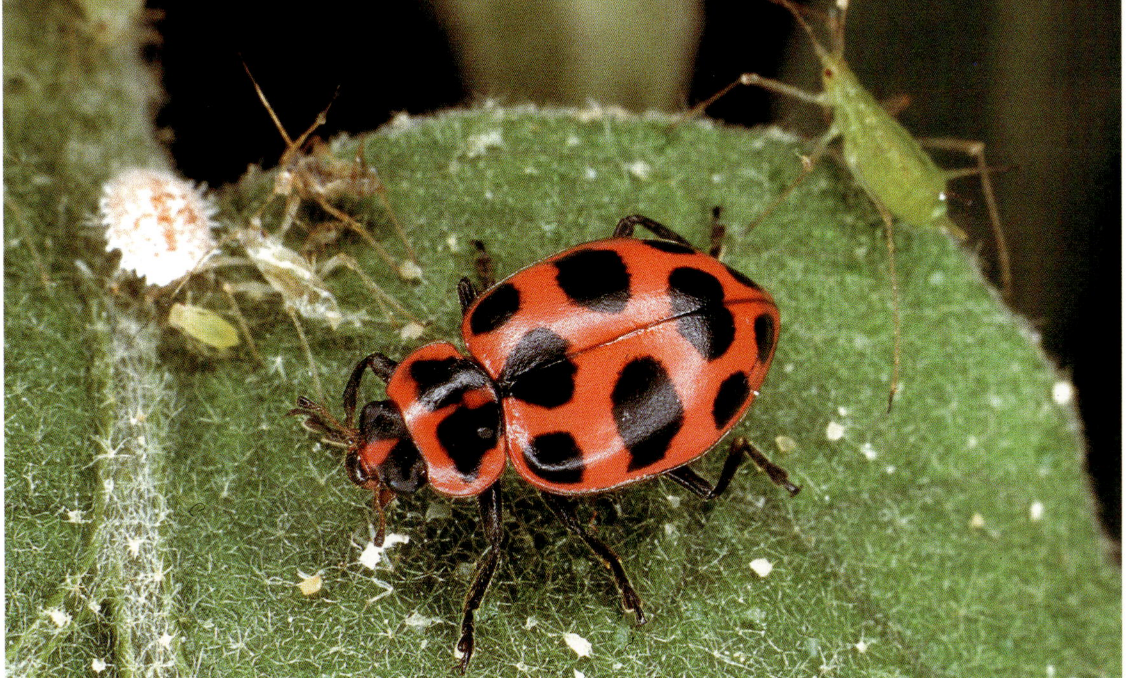

Ladybugs are treated tenderly in many different cultures.

Above: The pink spotted lady beetle *(Coleomegilla maculata)* surrounded by aphids, its favorite prey.

role in gardens than from the fact that the most widespread European species *(Coccinella septempunctata)* has elytra (hard outer wings) marked with seven spots—a number that has mystical connotations.

Crickets, cicadas, and praying mantises are foremost among the insects popularized and depicted in fairy tales and fables. In the West, these and other insects show up more frequently in literature than in art, whereas in China and Japan, they play a more important role in both poetry and painting. I was surprised to learn how much children in Japan cherish insects, which they tend carefully and often keep in small cages, sometimes slung across their shoulders. Their menageries may include cicadas *(semi* in Japanese), stag beetles *(kuwagata),* and rhinoceros beetles (subfamily Dynastinae), known as "helmeted insects" *(kabuto mushi).* Shintoism, which encourages respect for life in all its forms, plays an important part in the Japanese appreciation of insects. I doubt that any Japanese mother forbids her child to play with an insect because it is "dirty." (I am glad that my parents had a similar view of insects, even if they had no knowledge of Japan or Shintoism.)

Considering the millions of insects that people thoughtlessly squash to death every day, any accolades bestowed on insects are the exception to the rule. Although we honor some of the varieties just mentioned, humankind overall does not number insects among its trusted friends.

Based on only partial and fleeting observations, we often imagine the worst of insects that have even slightly unusual characteristics: spots resembling eyes, horn-like outgrowths, sophisticated mimicry and camouflage skills so clever that they could be

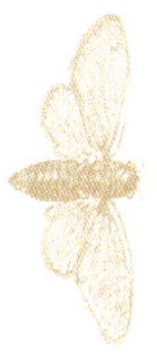

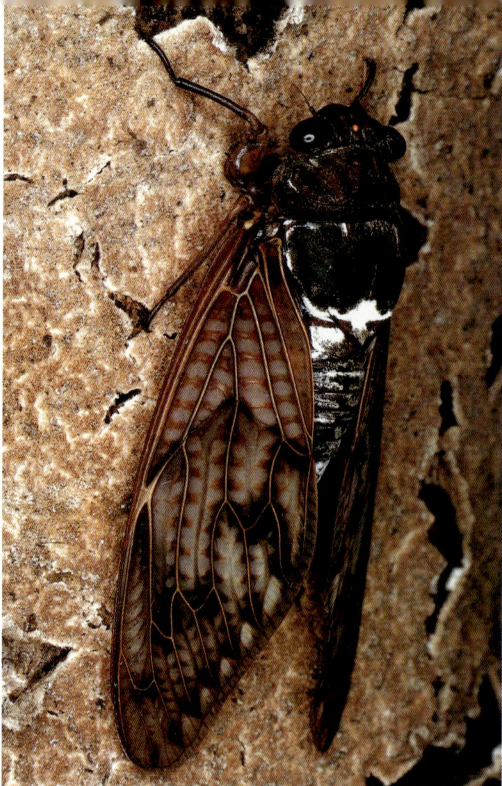

In Eastern cultures, certain insects have been widely represented in the visual arts and in poetry. The Western world has kept insects at more of a distance, and has never taken these creatures seriously.

Above: A cicada from Japan.

Below: A praying mantis from the Dominican Republic.

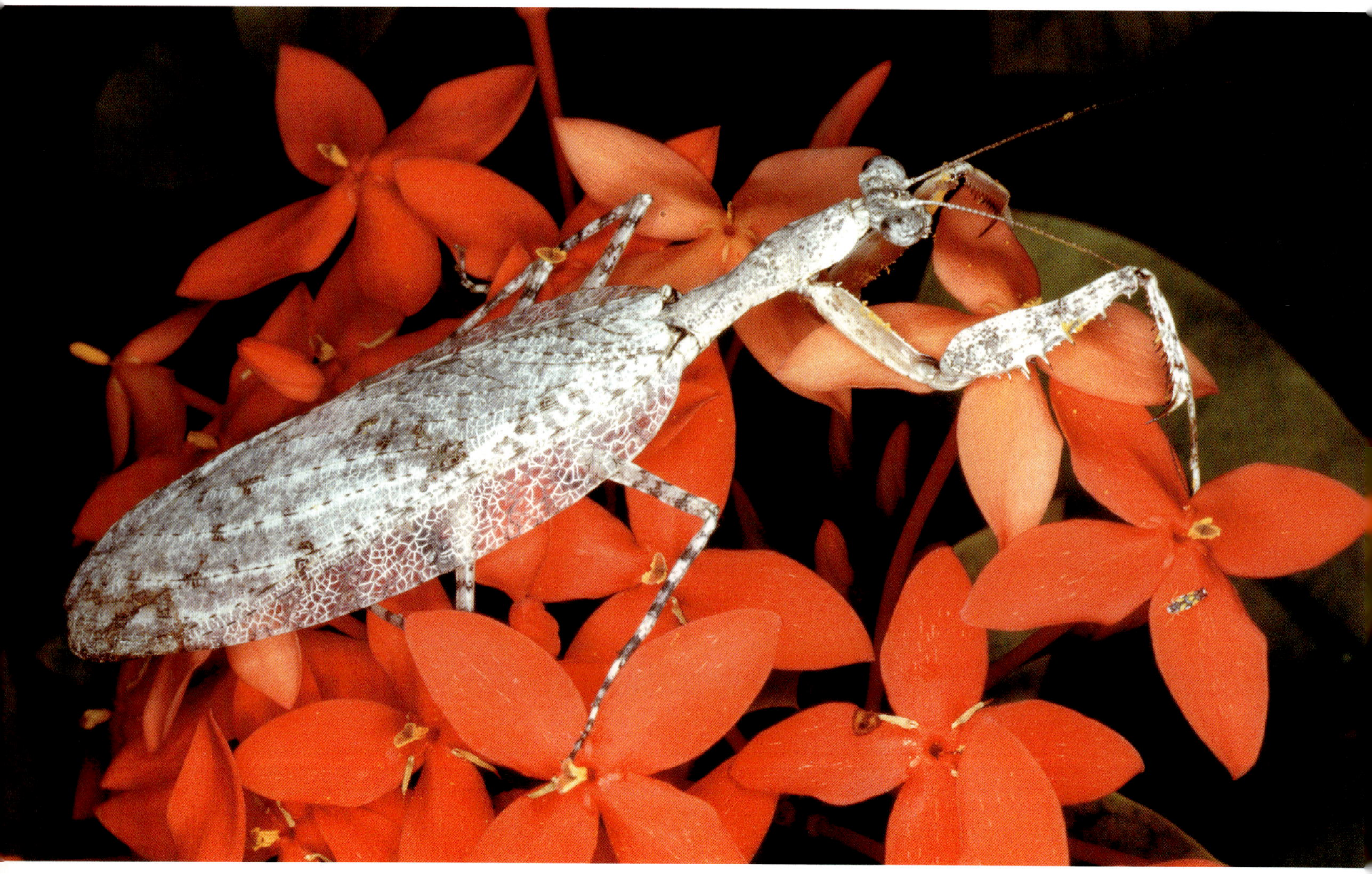

the work of some evil genie, and countless other peculiarities. Our storytellers latch onto these odd traits, turning innocent insects into fearsome villains with sinister intentions.

I once showed a Colombian woman various specimens that I had collected in her country. She was completely unfamiliar with these insects and pointed at the most suspicious-looking one, asking, "Is this one poisonous?" "Not at all," I replied. "What about that one?" "No, it's quite harmless." "Well, surely that other one must be."

One of them had to be guilty!

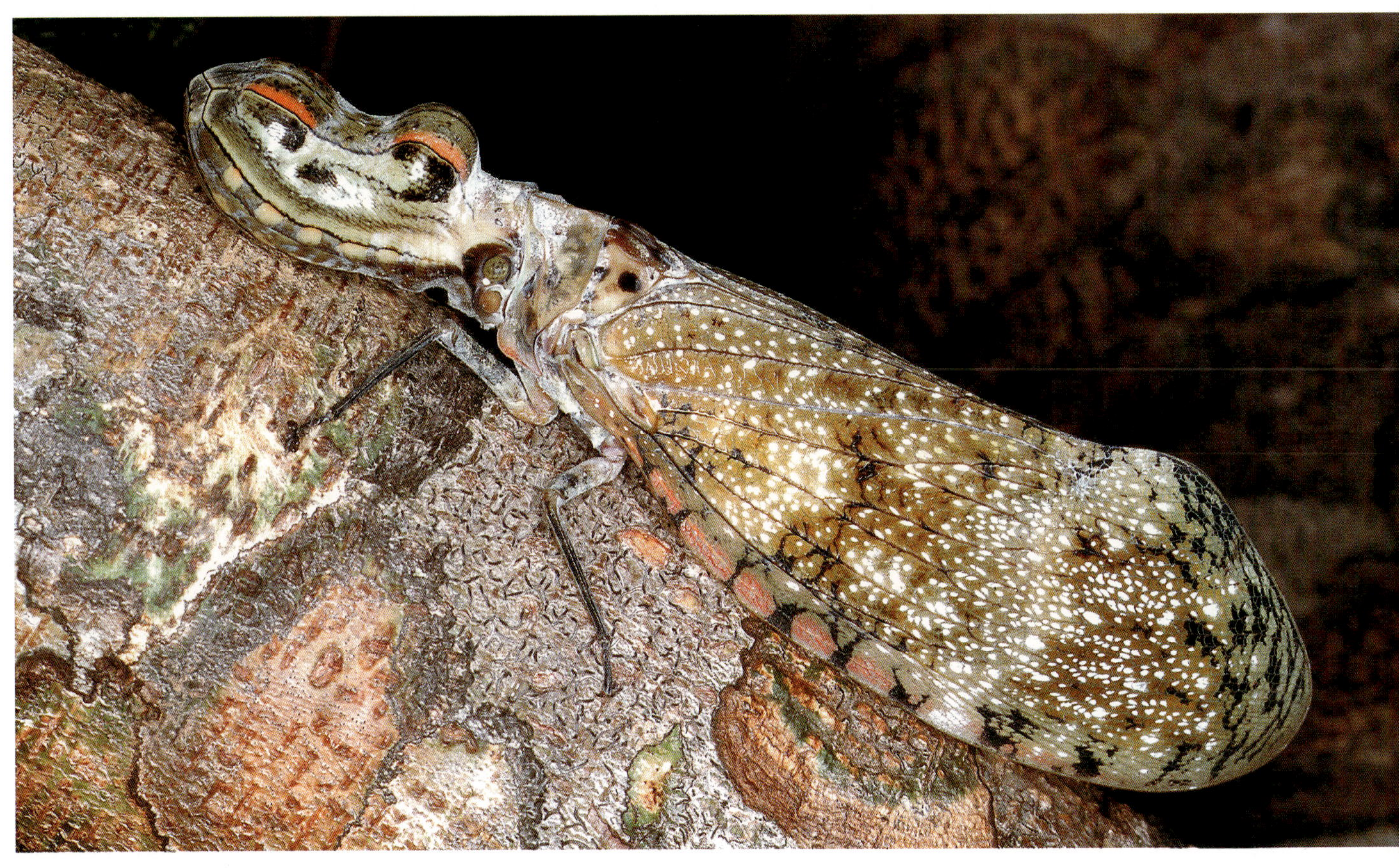

This large lanternfly (*Fulgora laternaria*, family Fulgoridae, order Homoptera) is from tropical America and has a wingspan of 15 cm. (6 in.). One of nature's little jokes, this bizarre insect often frightens people and has been dubbed the "viper cicada." Like its close cousins the cicadas, it has mouth parts that enable it to pierce plants and suck sap from them. Its unusual appearance may explain its bad reputation. It is humorously nicknamed the "peanut-head bug" or "flying alligator."

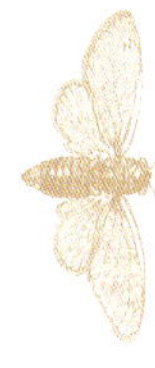

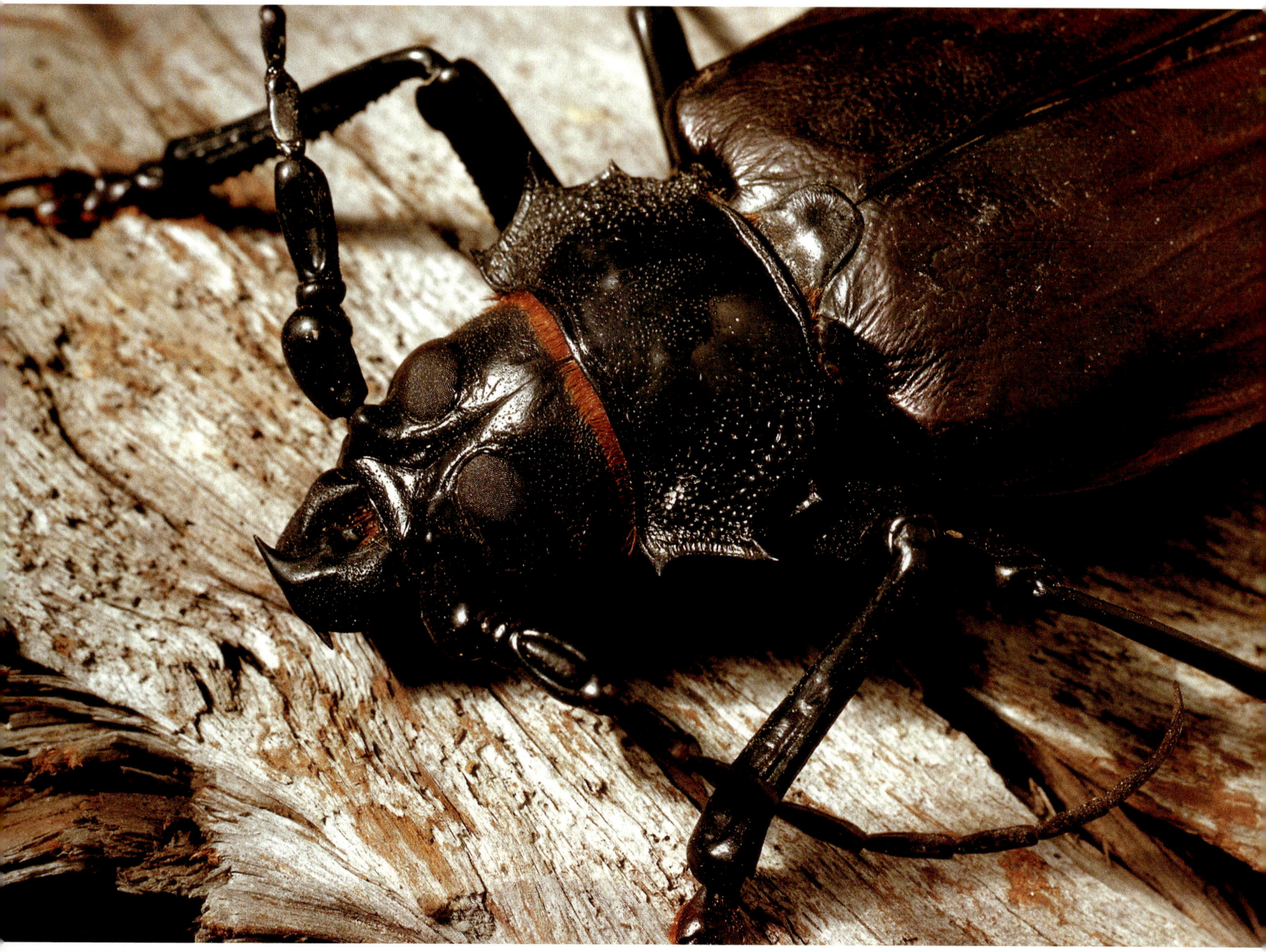

Faced with nature's astonishing diversity, we may find overly formalistic notions of beauty insufficient. However, if we embrace the notion of fascination, we may come to appreciate many bizarre, unusual, or extreme phenomena as beautiful in their own right.

A force of nature, the beetle *Titanus giganteus* (above) is certainly not beautiful in the same way as the gracious ebony jewelwing (*Calopteryx maculata*; opposite page). But seeing one of the giant beetles from Amazonia is as extraordinary as coming face-to-face with a jaguar or an anaconda. Maximum length: 17 cm. (6.5 in.).

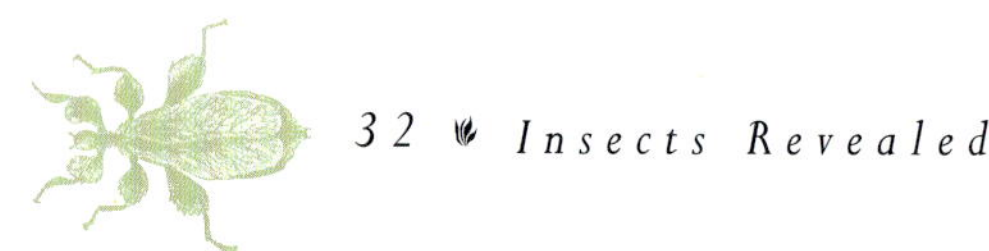

2

fascination: an expanded concept of beauty

The objective of this book is to reveal the beauty of insects to those who are receptive to it. I fear that my work is cut out for me: as already mentioned, insects have never been humankind's best friends, and any admiration for them remains selective at best.

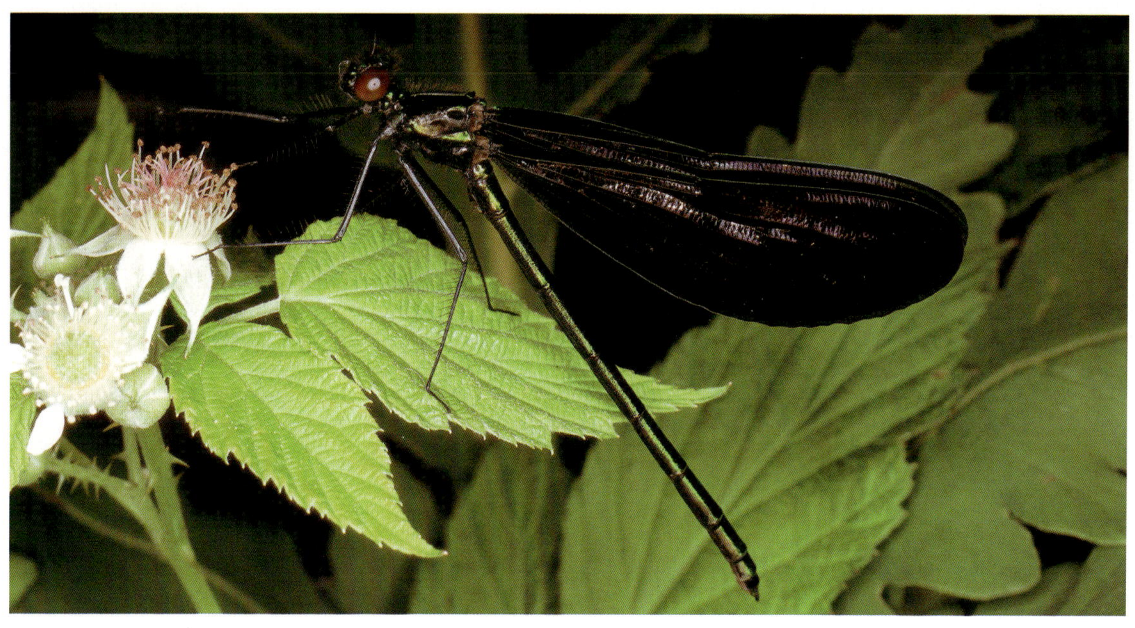

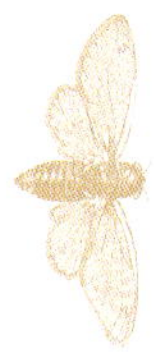

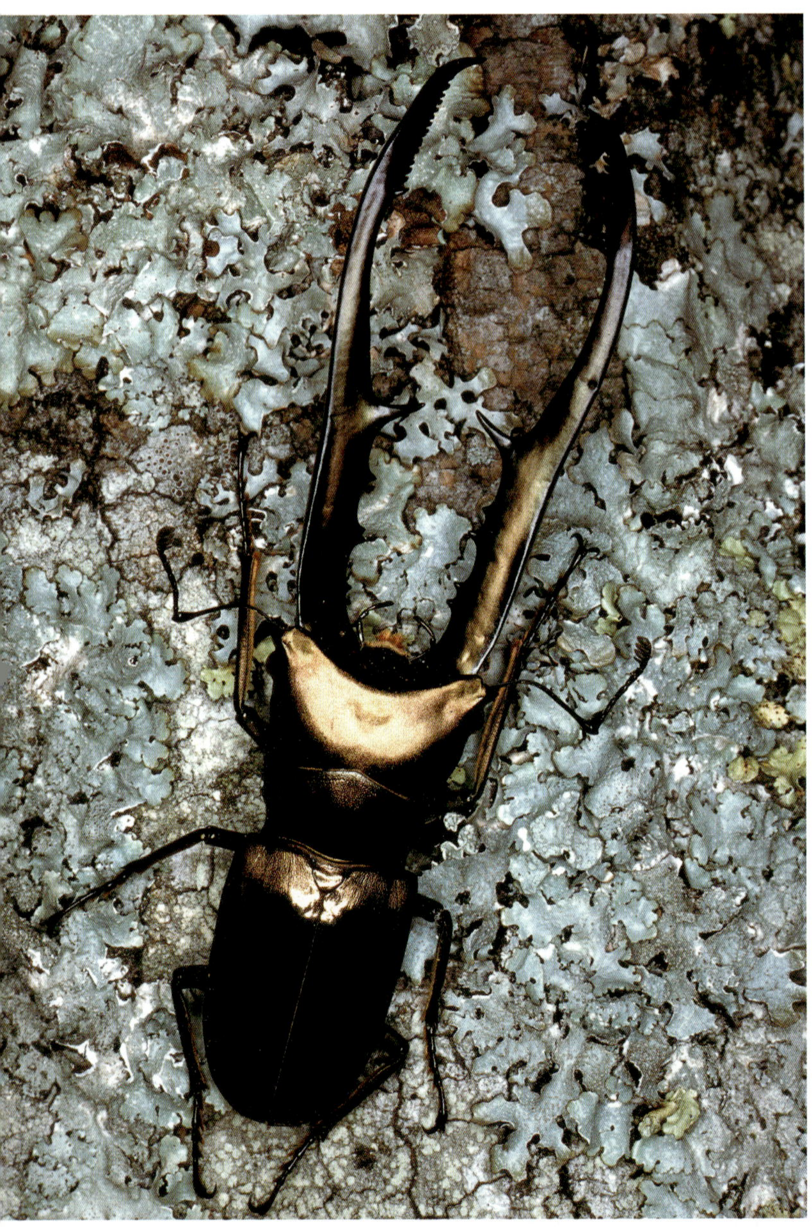

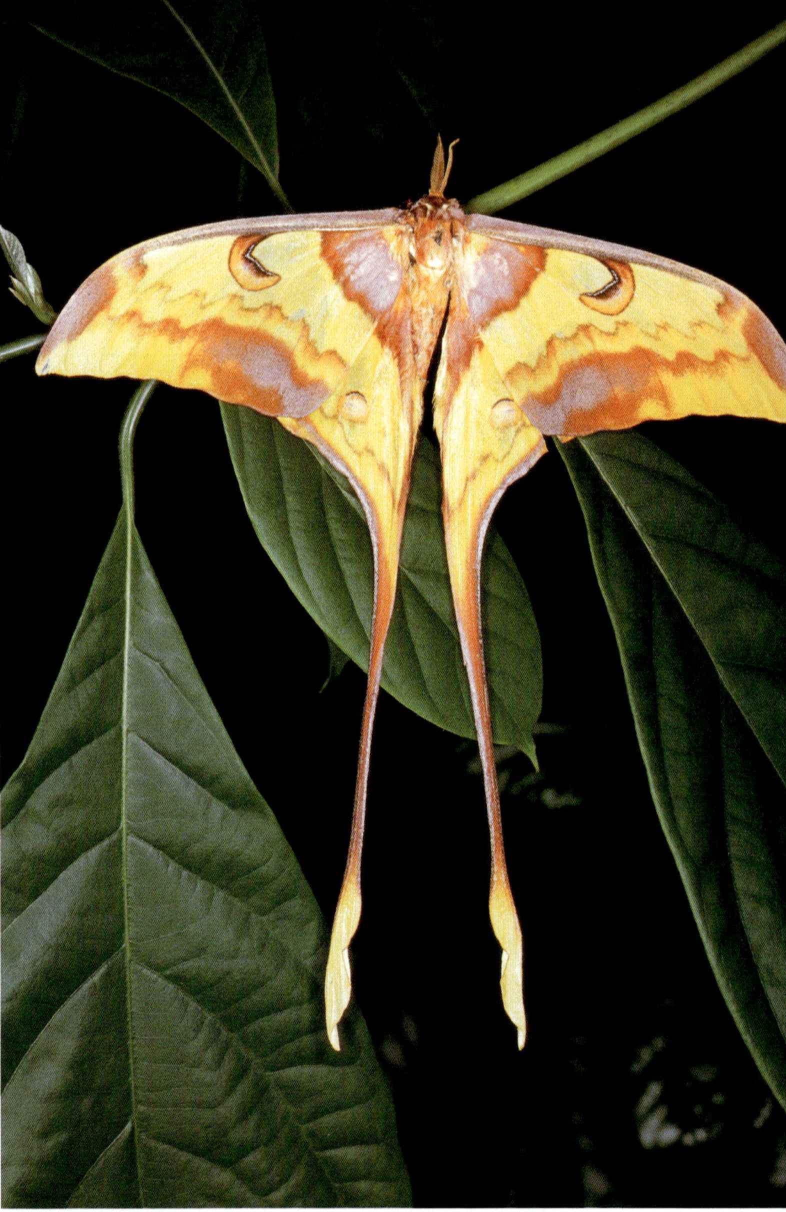

Nature in the tropics often assumes forms that may seem excessive to us. Nevertheless, balance is always maintained—otherwise, how could such forms have survived until the present day?

Left: A male beetle (*Cyclommatus metallifer,* family Lucanidae) from Sulawesi Island in Indonesia. This superb insect is equipped with mandibles that are actually longer than its body. In the survival of the fittest, such extravagant features could be detrimental; they might win in terms of length but fail in terms of functionality. Strong mandibles, however, prove essential during male-to-male combat for females.

Right: This large male luna moth (*Actias maenas,* family Saturniidae) from Southeast Asia has less to fear concerning the possible liability of the extremely long tails on its posterior wings, since these tails do not have the same vital importance as do male lucanids' strong mandibles.

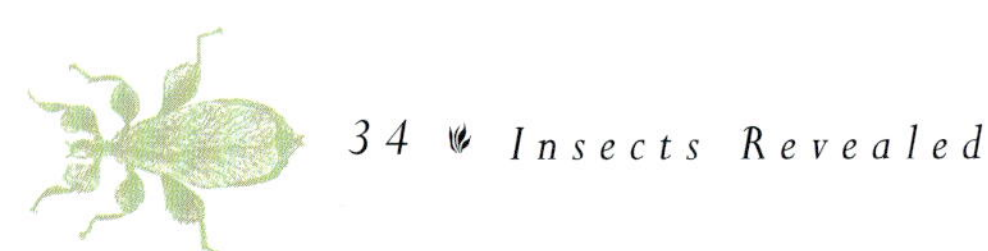

It is easy to extol the virtues of butterflies. However, does "beauty" apply to those insects that, to the public, often seem more like monsters than marvels? Can we truly speak of beauty in the presence of *Titanus giganteus* (one of the world's largest beetles) or the caterpillar form of *Citheronia regalis,* the regal moth? These creatures may well be impressive, but can one consider them beautiful? One cannot use the word *beauty* indiscriminately, as one might in speaking of flowers, for the more than one million heterogeneous, bizarre, unusual, or even frightening species of insects. Although they may be impressive, many insects will not appear beautiful at first sight. In numerous cases, a different term would seem more fitting, one that better describes a phenomenon that transcends visual appeal and is capable of touching us on a more visceral level: the words *fascination* and *fascinating* come to mind.

The notion of fascination helps us to develop an expanded conception of beauty, one comprising various species for which the term *beautiful* (in the sense of "pretty") might not apply. After having perused these pages, the reader might come to look kindly on the mask of *Petrognatha gigas* (a giant horned beetle) and exclaim, "What a magnificent monster!" A feeling of fascination may often come before any appreciation of beauty and may offer us glimpses into forms of beauty that remain unclassified (and unendorsed) by our esthetic codes.

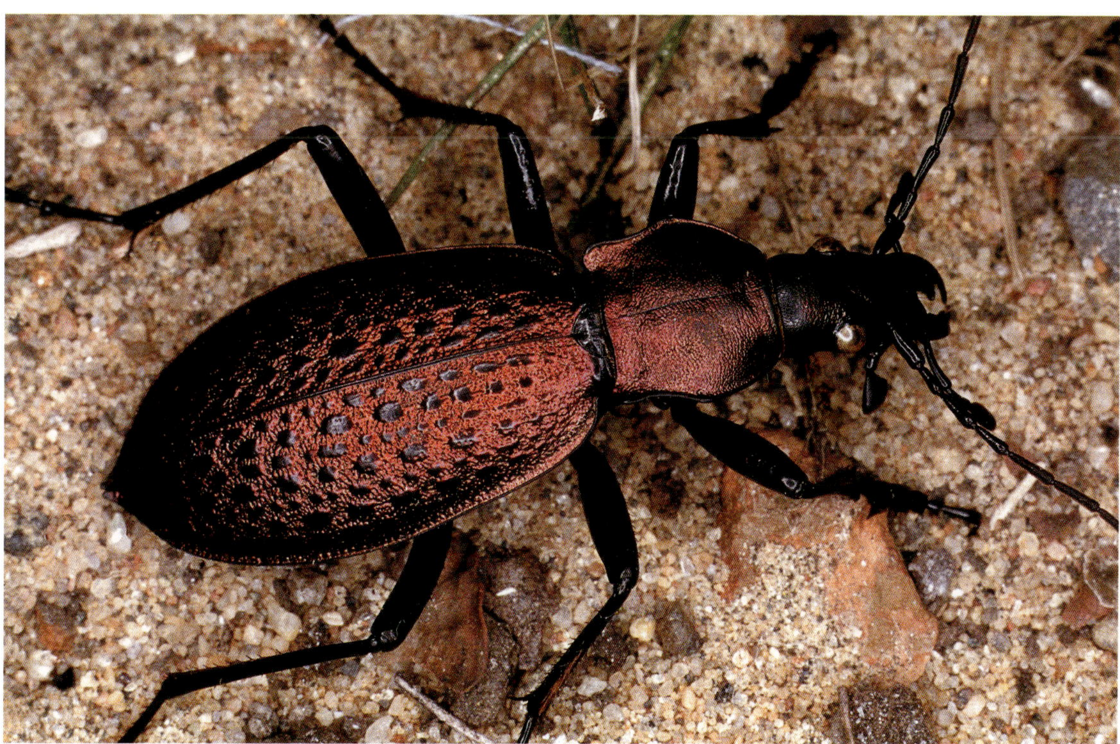

A carabid beetle from North Korea *(Coptolabrus smaragdinus).*

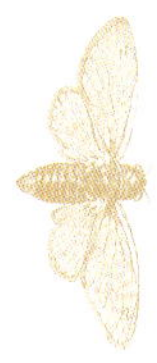

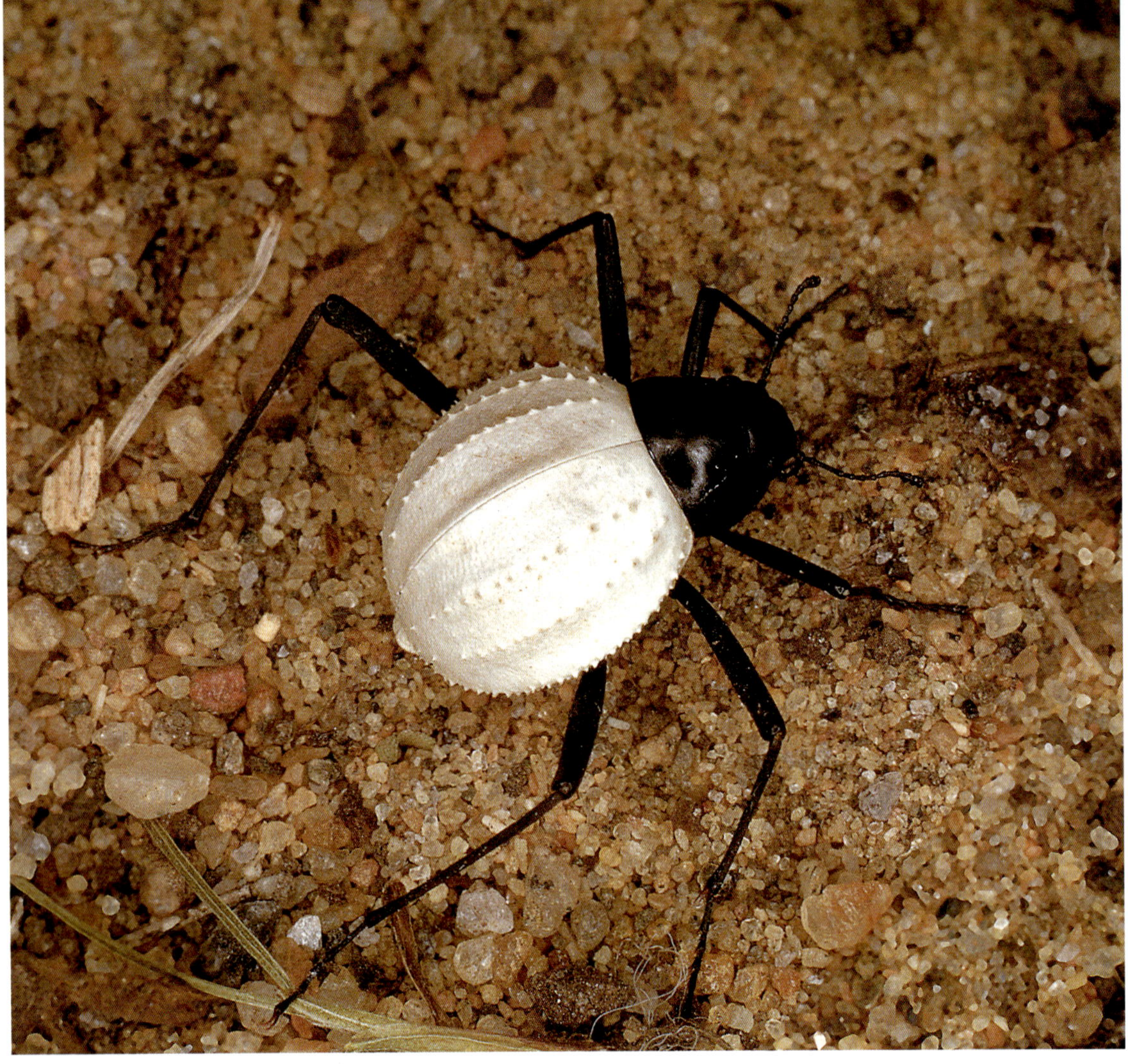

Despite its humble beauty, this small desert beetle, *Onymacris* sp. (family Tenebrionidae), is captivating nonetheless. It inhabits the Namib Desert in western Africa.

We are thus entering a world in which the quest for beauty has instead become a quest for fascination, which, in turn, rests on an even more elusive and intrinsic concept: the deeper meaning of all life forms, whether produced in nature or in art.

The notion of meaning thus takes precedence over our conceptions of beauty; indeed, meaning is a prerequisite for beauty.

The great American biologist Edward O. Wilson offers a one-word response to this question of meaning: Wilson coined the term *biophilia,* which is also the title of one of his books. What is biophilia? As living beings, we respond to life in any form, which we recognize spontaneously through the gift of empathy. Nature in its entirety thus becomes meaningful, fascinating, and, by extension, beautiful.

The subtitle to this book, *Monsters or Marvels?* may send even skeptical readers scurrying through meadows, woodlands, or other biotopes in order to take up this question of beauty in their own way. One might begin with some small friendly insect, such as a tiny green weevil with pink-tinged wings. The gift of biophilia, which we all share, will help the reader to complete the journey.

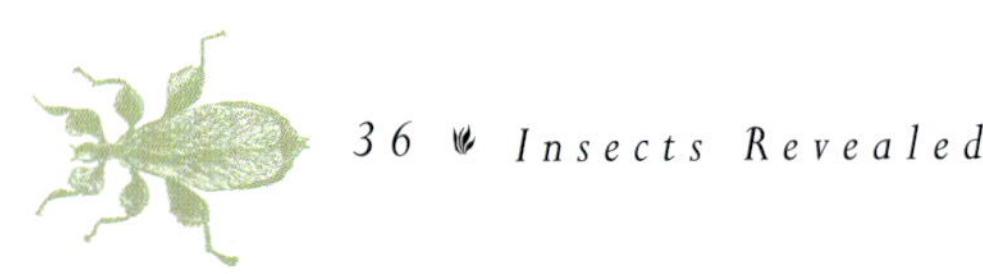

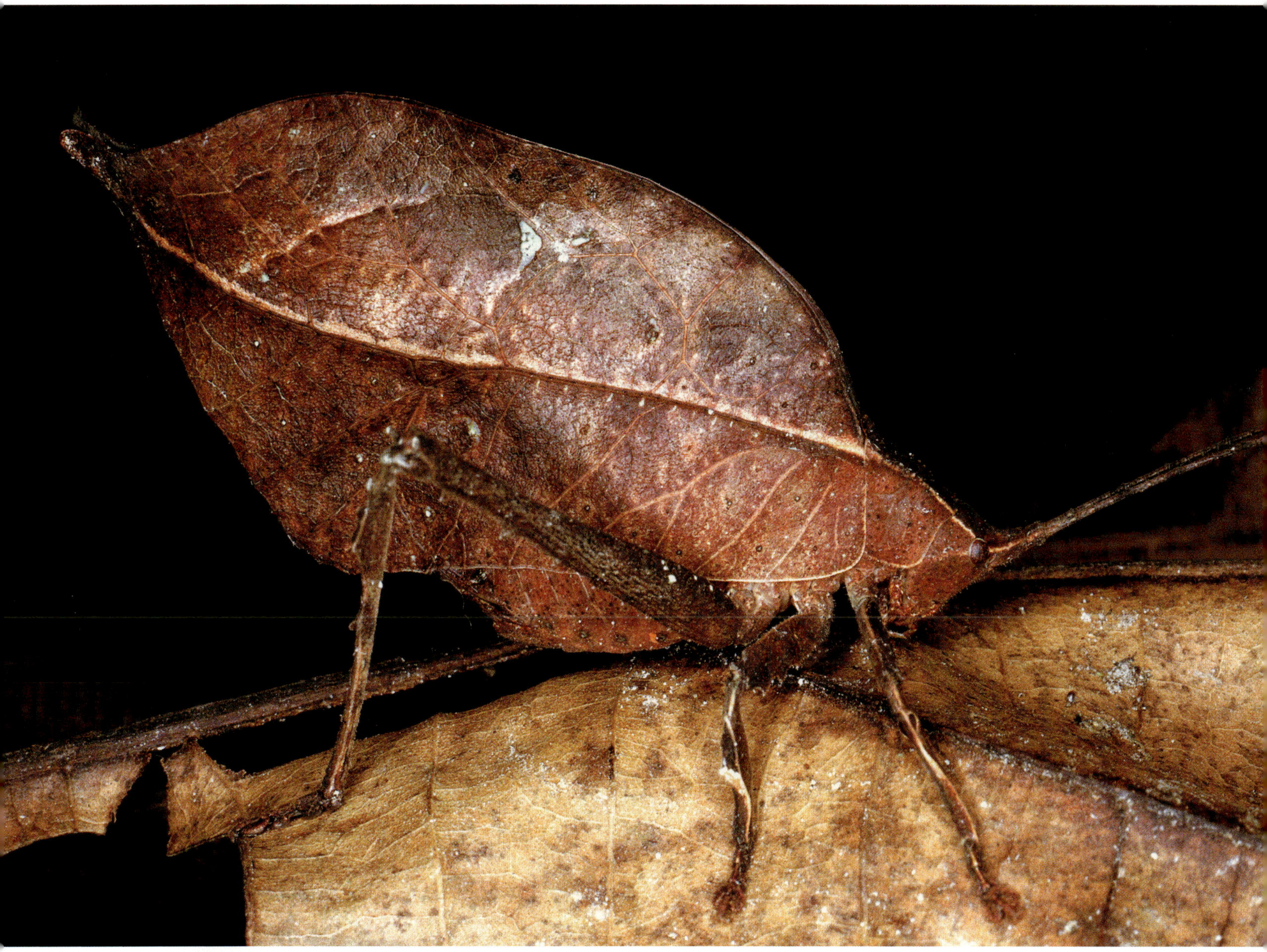

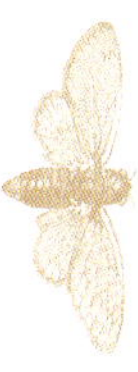

A leaf grasshopper from the Amazonian region of Peru. Unlike its brightly colored brethren, this grasshopper, along with most insects, adopts dissimulative coloration and behavior. These species have a beauty that appeals less to the eye than to the mind that explores nature's mysteries.

As is the case throughout the animal kingdom, each species has its own personality. Some are flamboyant, powerful, and belligerent; others are humble and gentle, and exhibit a tender charm.

Once embarked on this journey, the reader will also be able to appreciate a facet of insects' lives that photography, at least as I practice it, is unable to express: the qualities of motion and animation that transcend insects' forms and enhance their beauty. Yet even if our hearts skip a beat when we observe insects flying or rushing around at top speeds, such experiences are fleeting and leave us no time to become properly acquainted. So I take consolation in this thought: even if I cannot convey the sense of movement that accounts for a large part of the magic of insects—and of all living things—I can at least communicate their forms, colors, textures, and habits. These qualities can only be appreciated through calm and prolonged observation.

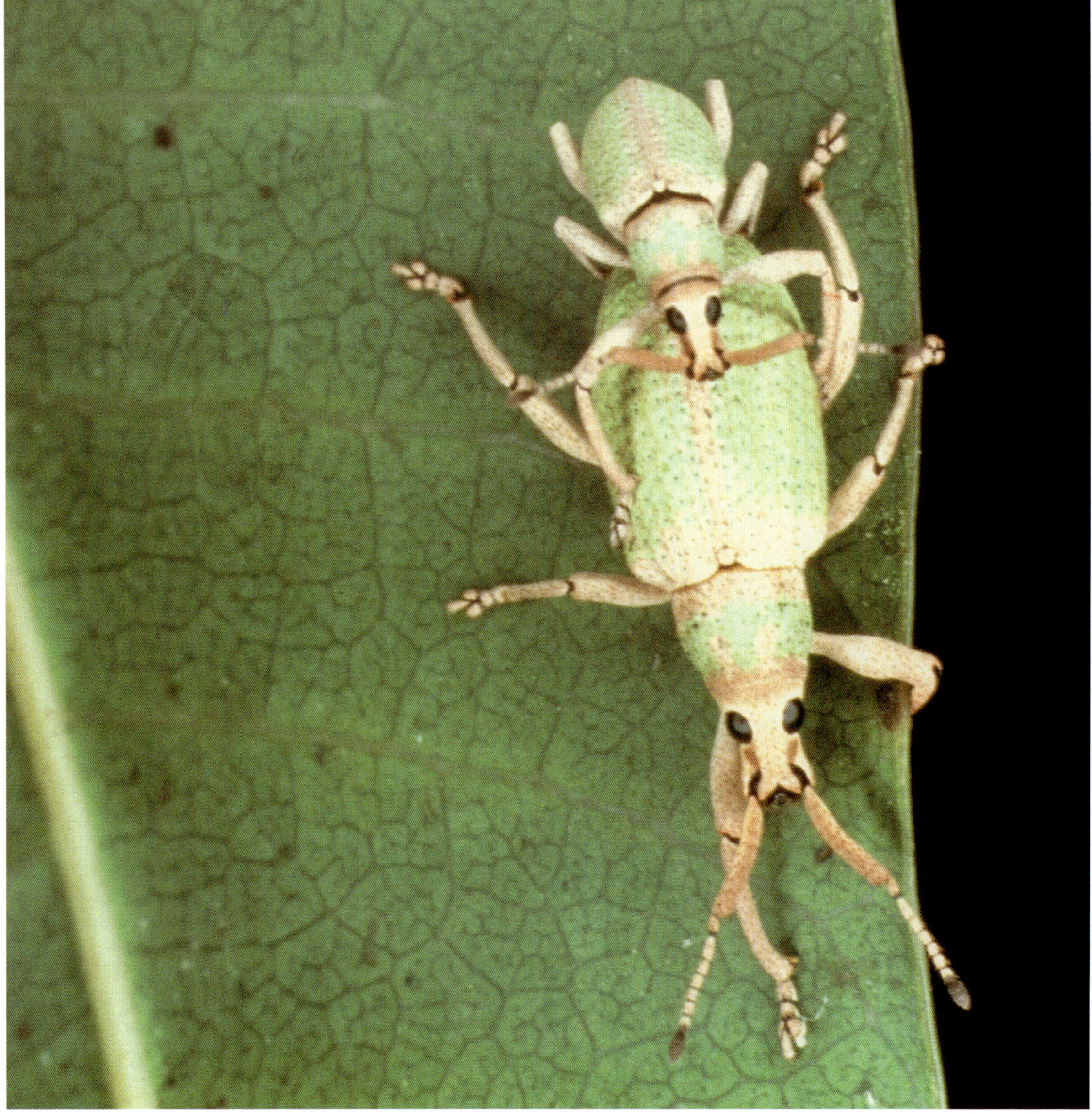

Weevils mating, Trinidad.

the classification of
living things:
an intellectual need

As they observed the world around them, the first hominids—members of the primate superfamily, including the anthropoid apes and man—must have established fundamental distinctions based on which organisms were edible, which ones stung, and other characteristics related to survival.

The initial rudiments of classification were undoubtedly similar for our hunting ancestors and for the creatures around them, at least until the invention of language. Then it became possible not only to re-experience sensations, but also to evaluate and appreciate the qualities of things by naming them. Things could be stored in memory, from which they could be recalled by speaking their names. More important, words enabled our ancestors to relate their experiences and to develop a store of knowledge that could be handed down to future generations.

However, progress in this regard seems not to have been particularly rapid, since the Greek philosopher Aristotle was the first to extend this naming process beyond life's practical requirements.

Aristotle decided to compile an inventory of the animal kingdom in which each organism would be classified according to its similarities to and differences from other organisms. He spent three years on his self-appointed task, which remained incom-

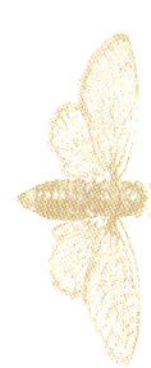

plete owing to the extent of the undertaking and the shortage of biological material from regions other than the Mediterranean basin.

Until the Renaissance, people in the Christian West explained the natural world mainly in supernatural terms and did not seek to enumerate its contents. Humans were only temporary inhabitants of this world, and their final destination, heaven, required their undivided attention. Medieval Christians did not put an emphasis on observing or understanding the world, from which they ultimately had to detach themselves; they imagined the Earth as a stage upon which the forces of good and evil, embodied in all sorts of natural objects, were in constant conflict, and this made it difficult to distinguish between the natural and the supernatural.

Over the next several centuries, science made some advances thanks to the efforts of a few pioneering individuals, but a universal classification system for living things—one that would provide a comprehensive and intelligible picture of reality—remained an elusive goal. Credit for this achievement goes to the eighteenth-century Swedish botanist Carolus Linnaeus.

Carolus Linnaeus (1707–1778)

A trained physician with a passion for botany, the great Swedish naturalist Linnaeus worked with a large number of collaborators, recruited from among his students. Their passion was so great that they were known as "Linnaeus's apostles"; they traveled to the four corners of the earth to collect as many specimens as they could for their master. Linnaeus described and classified 7,700 plant species and 4,400 animal species, including many insects.

The taxonomic system he set forth in *Systema naturae* (the first edition appeared in 1735), had the following objectives:

—to create a system that could be used to classify every organism, known or as yet unknown;

—to achieve international recognition for the system;

—to make the system internationally accessible by using Latin, a transnational language that all scientists understood (Latin had already been modified in the Middle Ages, and Linnaeus continued this modernization process by Latinizing proper or geographical names that were new to the language).

Modern taxonomy, the science of classification, still uses the Linnean system. The founding principles were so well chosen and so flexible that its general structure was able to accommodate numerous modifications and additions over the years; many previous taxonomic systems had failed this test.

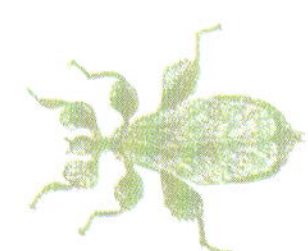

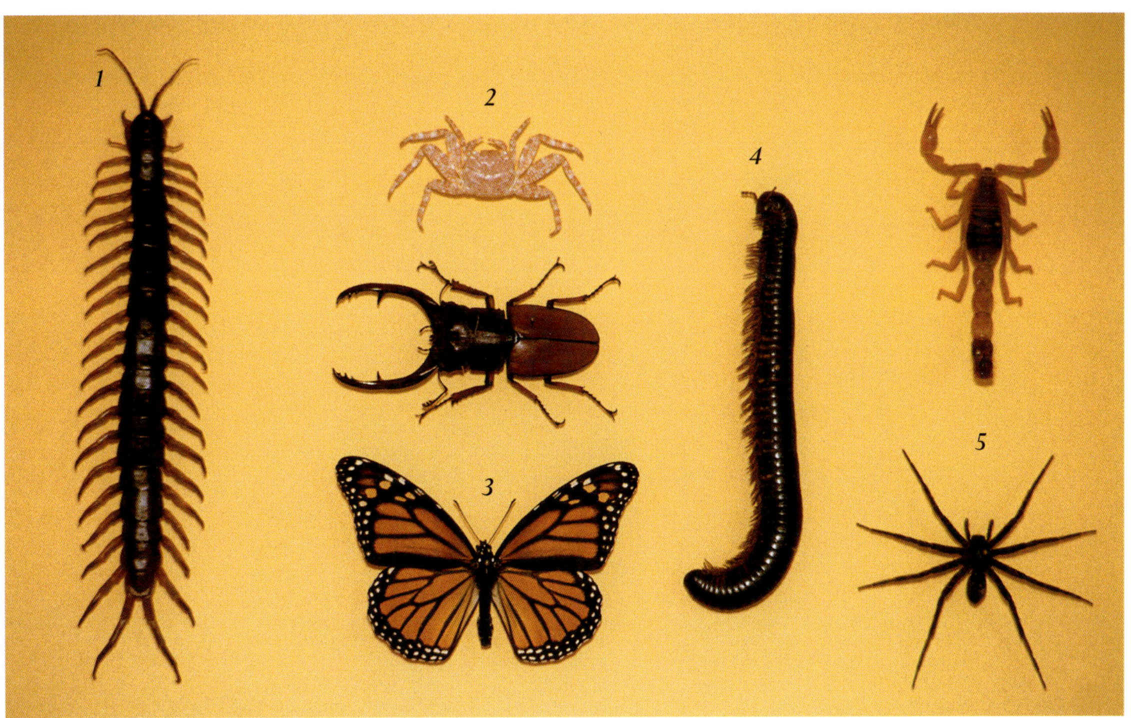

The arthropod phylum includes four main groups:

Myriapods: centipedes and millipedes (1, 4)

Crustaceans: crabs, lobsters, and shrimp (2)

Insects: beetles, butterflies (3)

Arachnids: scorpions and spiders (5)

An overview of the Linnean system and its classification of insects

The many individual species that make up the animal kingdom are divided into several branches, or phyla; one of these, the arthropods, comprises invertebrate organisms whose bodies are made up of articulated segments, similar to a suit of armor. This "armor," or exoskeleton, is the common trait of all arthropods.

Four major groups within the arthropod phylum are listed below, with some of their representatives:

Myriapods: millipedes and centipedes

Crustaceans: lobsters, crabs, and shrimps

Arachnids: spiders, scorpions, and horseshoe crabs

Insects: butterflies, dragonflies, and junebugs

The class Insecta is divided into orders such as Lepidoptera, Coleoptera, and Diptera, which are differentiated primarily by their wing structure (the suffix *-ptera* derives from the Greek *pteron,* wing).

The orders are grouped into suborders, families, genera, and species, which are sometimes further divided into subspecies and regional or seasonal varieties.

The names of insects and other animals and plants include both genus and species. The first part of the name (the generic) is capitalized and the second part (the specific) is written in lowercase, for example, *Titanus giganteus.*

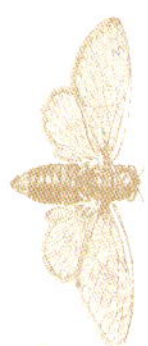

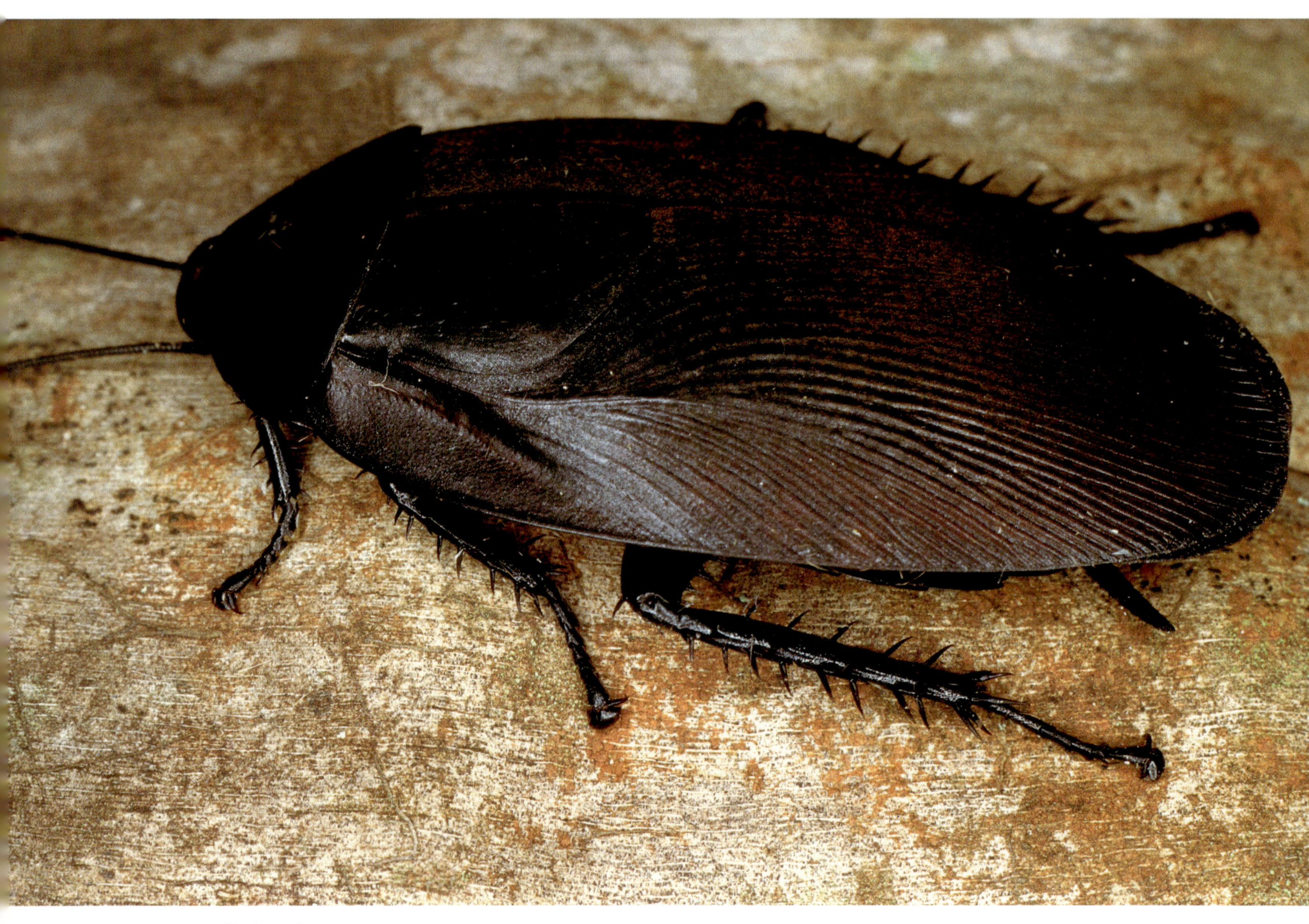

Cockroach.

4

the origins of insects

It may be difficult to imagine that familiar insects such as silverfish, dragonflies, cockroaches, and mayflies existed during the Carboniferous era some 300 million years ago, and in much the same form as today. One can surmise that the origins of their ancestors—vermiform (worm-like) marine invertebrates that became terrestrial—reach even farther back in geological time, most likely to the Precambrian era.

Since there are no fossil remains of these marine invertebrates, however, this supposition remains hypothetical.

It was long believed that the oldest insect fossil dated back 375 million years. However, the journal *Science* recently reported that a paleobotanist from the Smithsonian Institution working on Quebec's Gaspé Peninsula had discovered the fossil remains of a wingless insect from 390 million years ago. Although this discovery effectively extended the chronology by 15 million years, much still lies in the realm of conjecture and hypothesis.

Whatever the origins of this primitive marine insect, whether vermiform or not, it must have evolved and adapted gradually to an air-based existence, at least in its adult stage. The larvae of virtually all insects that arose during the Carboniferous era are aquatic, with the exception of cockroaches; this uniformity suggests that the very first insects had aquatic ancestors. A small number of terrestrial and aerial insects of more recent origin have returned to an aquatic environment during either their larval

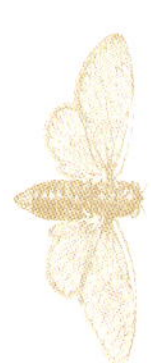

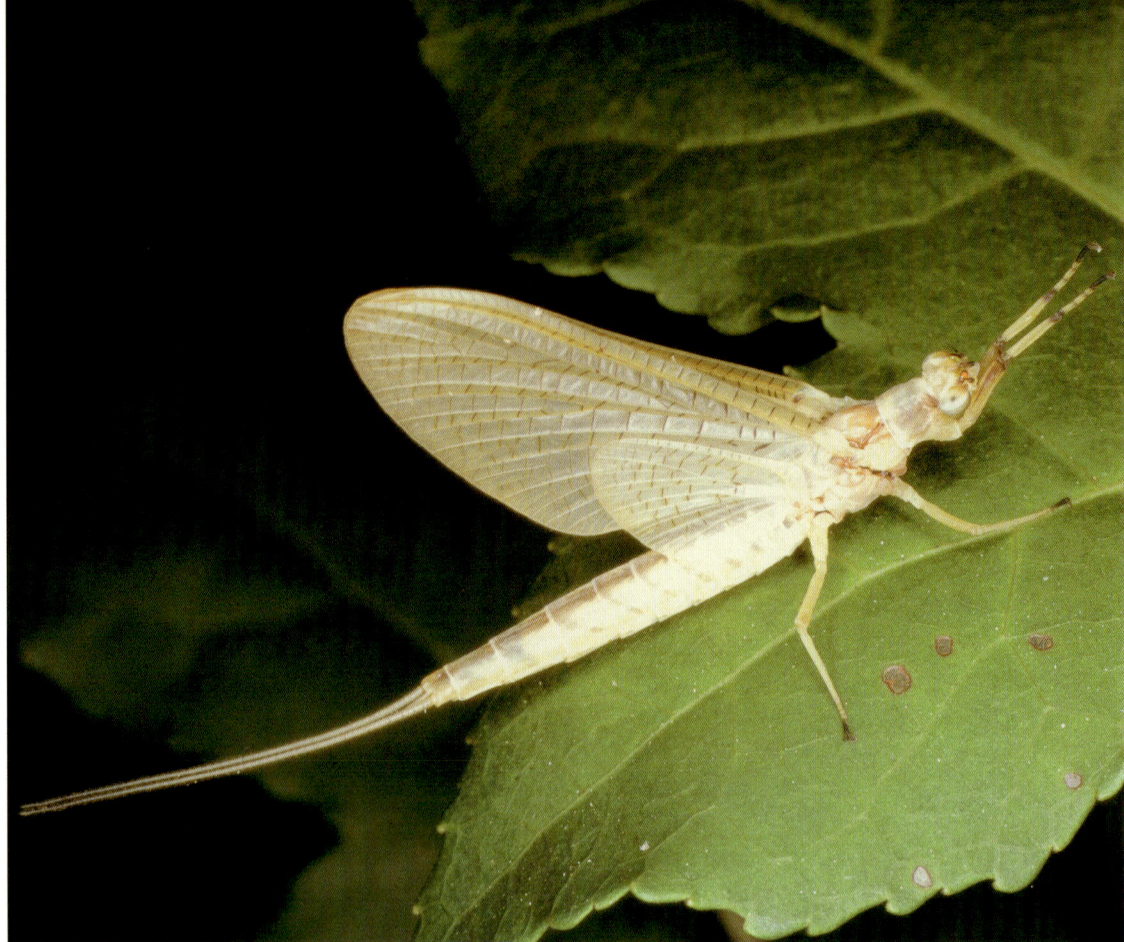

Mayfly (*Hexagenia* sp.).

or adult stages; these insects have developed a variety of adaptations that allow them to survive under water. Since they never developed gills, adult aquatic beetles such as the dytiscids (diving beetles) and hydrophilids (water scavenger beetles), and aquatic Hemiptera (true bugs) such as notonectids (backswimmers), ranatra (waterbugs), and nepids (water scorpions), use diverse means to transport air bubbles underwater. Others have developed a siphon-like structure at the end of their abdomens by which they replenish their air supply. In addition to these various strategies, these insects all possess spiracles, or breathing pores, that enable them to breathe as they fly from pond to pond.

Insects' adaptation to terrestrial life depended directly on the plant life that existed at the time. Jacques Brosse reminds us that "the rise of insects corresponds to the genesis of terrestrial vegetation. . . . The destiny of insects is governed by plants. Once plants conquered continents, insects followed."[3]

These primitive wingless organisms did have the characteristic six legs of the hexapods, but they had few other attributes of present-day insects. Consequently, although these pioneers of a long and continuous evolution deserve mention (their

3. Jacques Brosse, "L'insecte," in *Encyclopédie essentielle* (Paris: Robert Delpire, 1968).

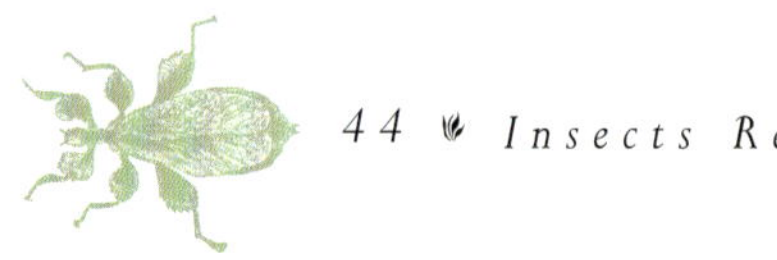

descendants today include the wingless insects of the order Thysanura), we will focus here on the winged hexapods (pterygots), which constitute what we call the "modern" insects.

The distinguishing feature of the pterygots (from the Greek word *pteron,* wing) is that they are equipped with wings. In addition, their bodies typically have three separate sections, and they have six legs.

Insects' wings are unique within the animal kingdom because they were not produced by the modification of another appendage such as the front legs, as is the case with birds or bats. The evolutionary progression of insects' wings is somewhat obscure. Some primitive insects had lobed outgrowths that served as rudimentary wings, perhaps cushioning the fall of those that attempted to fly. In all likelihood, these lobes evolved into the highly differentiated and specialized wings of the Lepidoptera (butterflies and moths), Coleoptera (beetles), Hymenoptera (bees, wasps, ants, and sawflies), and other flying insects. Irrespective of their evolutionary progression, these organs of flight proved essential to the dispersal of the insects since they made new food sources available. Until they had wings, insects fed on either plant life or vegetable matter decomposing on the damp ground.

Wings also contributed to the diversification of insect types, such as the huge dragonflies of the Carboniferous era—impressive aerial predators that patrolled the marshes in pursuit of other flying insects.

Long after the appearance of wings, another important development directly influenced the lives of insects and contributed to their further diversification: the gradual appearance of the flowering plants, which occurred from 150 to 70 million years ago. Thus began the interdependence between foraging insects and plants that required pollination. Early on, foraging insects located suitable plants by means of olfactory sensors; later, the development of color sensitivity reinforced this mutually beneficial relationship.

Ever since, most of the plants that make up our planet's cover of vegetation have depended on insect pollination for reproduction. Similarly, insects have come to rely on plants for food and shelter during different stages of their life cycles.

When humans became established as a well-defined animal species over one million years ago, our present-day insects had already existed for millions of years—with

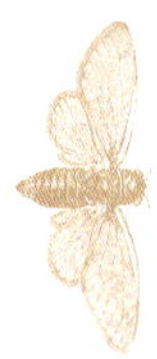

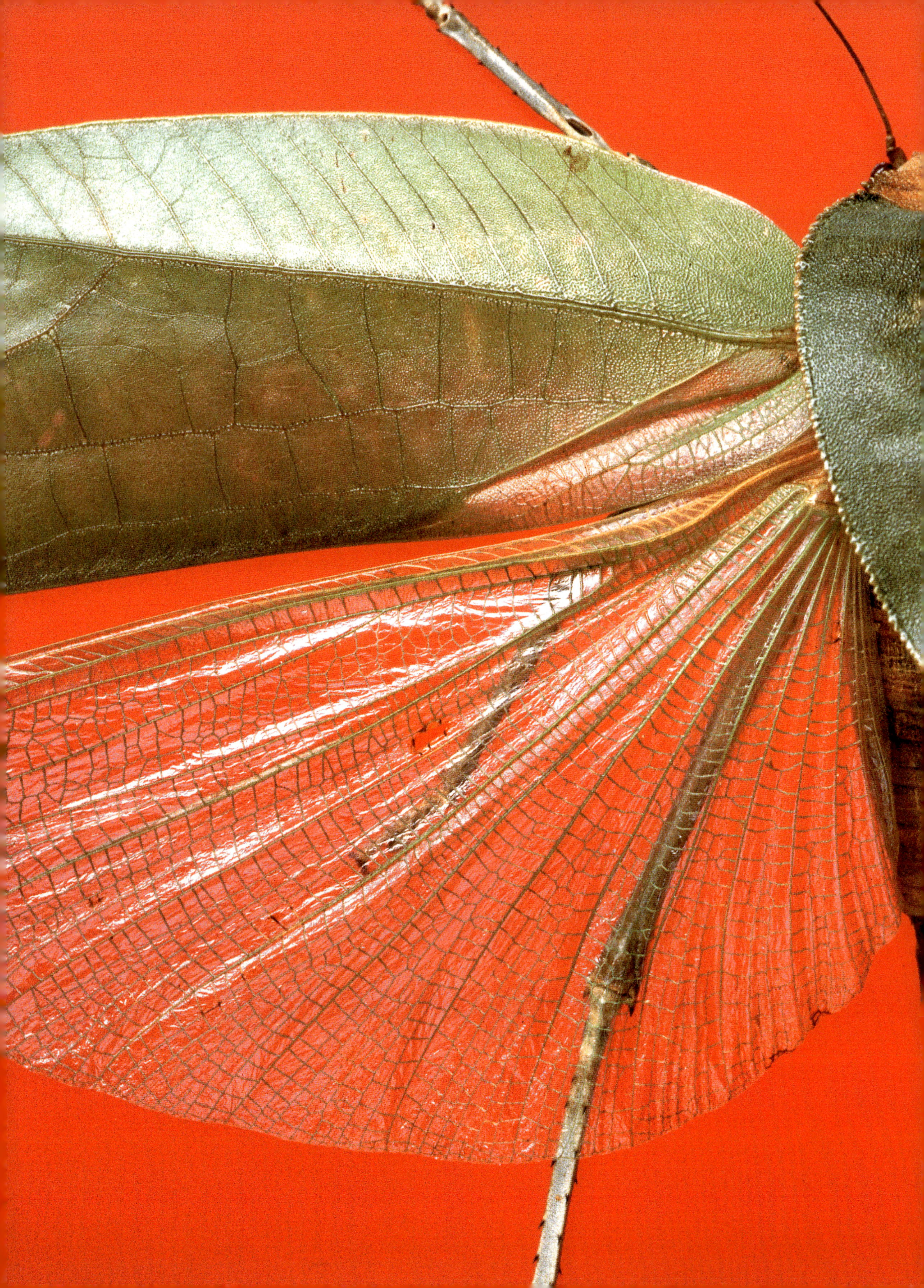

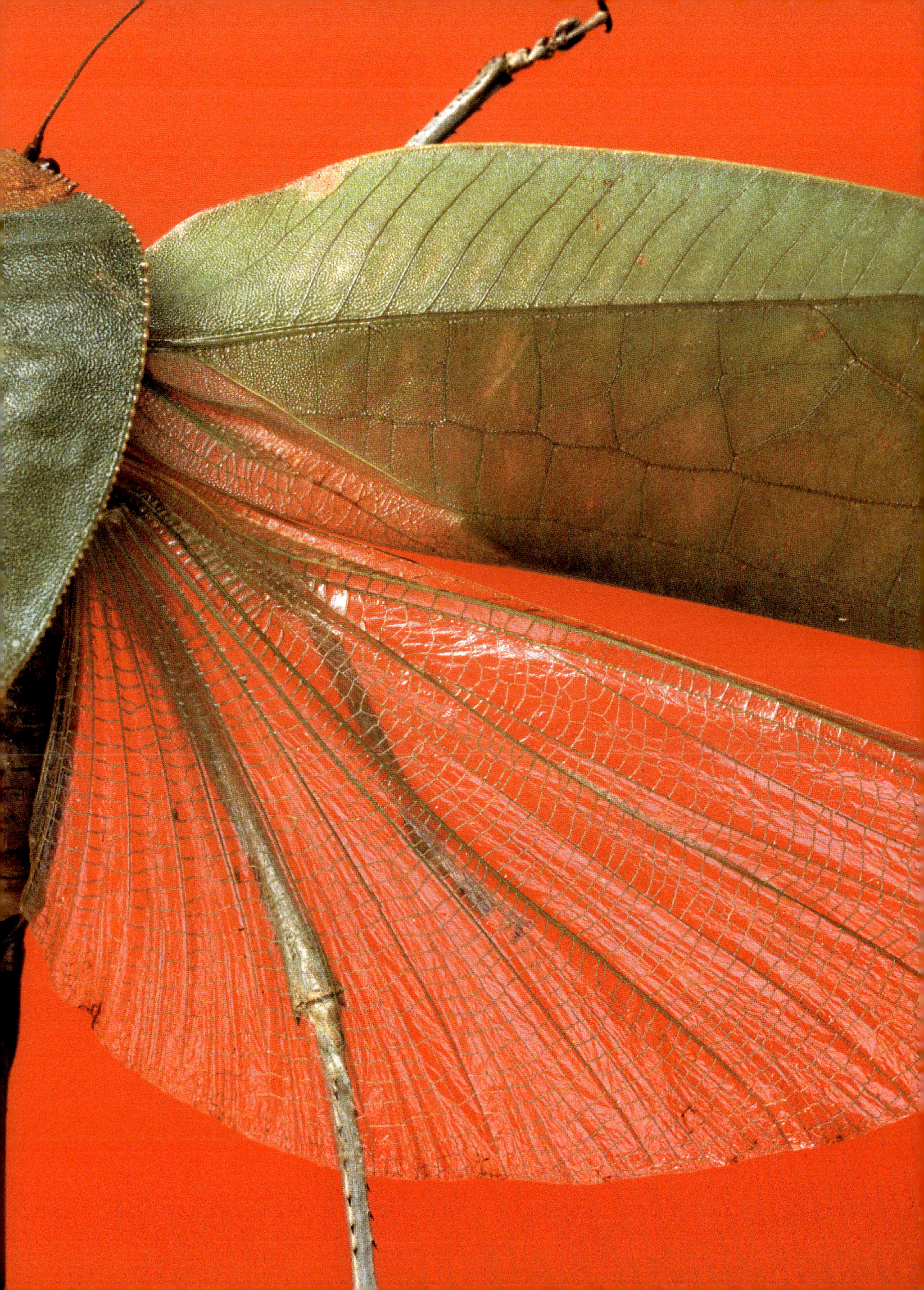

all the beauty and complexity of the specimens we capture today. Insects have thus witnessed much more than humankind; they were present for the arrival of the first amphibians, the dinosaurs, and all subsequent groups, including birds and mammals. Insects made their way across the continents as these land masses drifted through the oceans and endured earth-shaking cataclysms long before our ancestors rose up on their hind legs to become *Homo erectus* and then *Homo sapiens*. The premonitions in our legends are thus on the mark: insects really do come from another world.

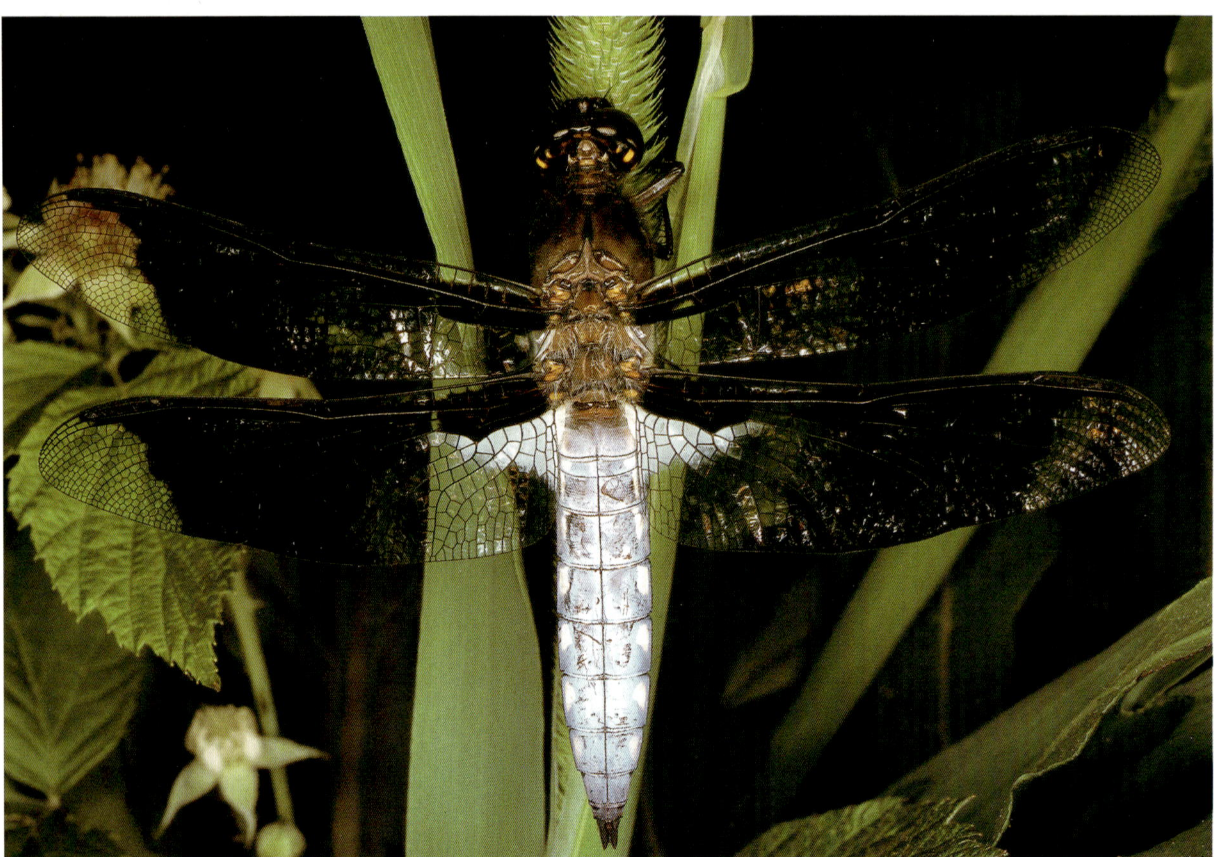

Mayflies, cockroaches, and dragonflies are among the oldest of the world's winged insects.

Above: Common whitetail dragonfly *(Plathemis lydia)*.

Preceding pages: This tropical leaf katydid (*Siliquofera grandis,* family Tettigoniidae) is found in New Guinea. The development of insects' wings represents one of the animal kingdom's truly original creations. More than any other factor, this locomotive apparatus has fostered the dispersal of insect species.

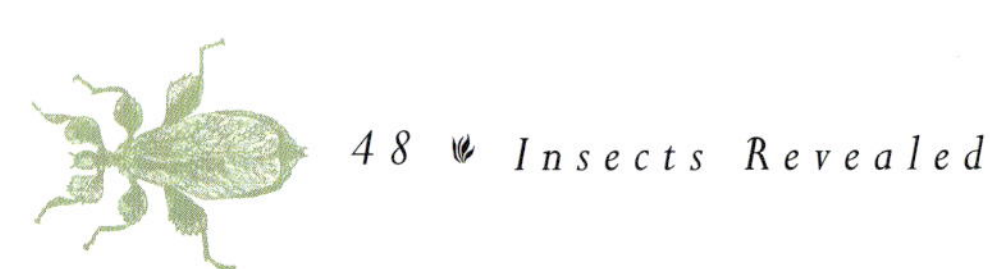

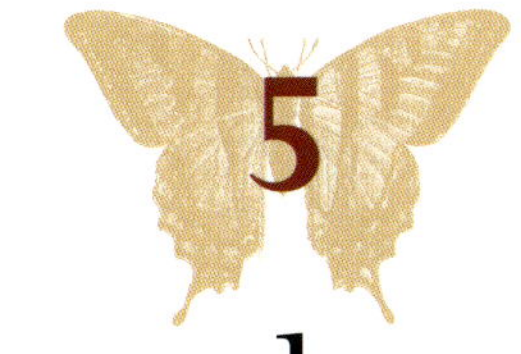

up close:
insect morphology

Unlike vertebrates, whose muscles and ligaments are attached to their internal skeleton, arthropods, including insects, have all of their anchoring points on the inner wall of their external skeleton. Arthropods consist of an aqueous mass within a segmented and fairly rigid structure, articulated much like a suit of armor.[4]

This external skeleton (or exoskeleton) derives its strength from chitin, a fibrous polysaccharide, and sclerotin, a protein that serves as a hardening agent. Sclerotin, which permeates the chitin in much the same way as plastic resin is combined with fiberglass, provides the body parts with varying degrees of hardness according to their functions. The abdominal sections of butterflies are quite soft, whereas the wings of certain beetles are extremely hard. Durability is also an essential feature of beetles' tarsal claws and mandibles, which must retain their sharpness and cutting ability throughout an insect's life—thus the question frequently asked when people examine a collection of beetles with brilliantly polished carapaces: "Are those made of plastic?"

---

4. Merely because insects are contained within this armor, one should not assume that they are imprisoned or somehow cut off from the outside world. The cuticle, the hardest layer of the integument, is covered with tactile hairs that provide insects with information on air currents, the relative positions of their body parts, and the nature of the surface they are resting on.

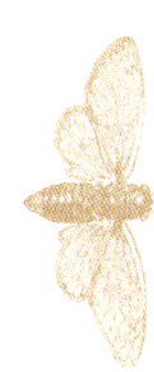

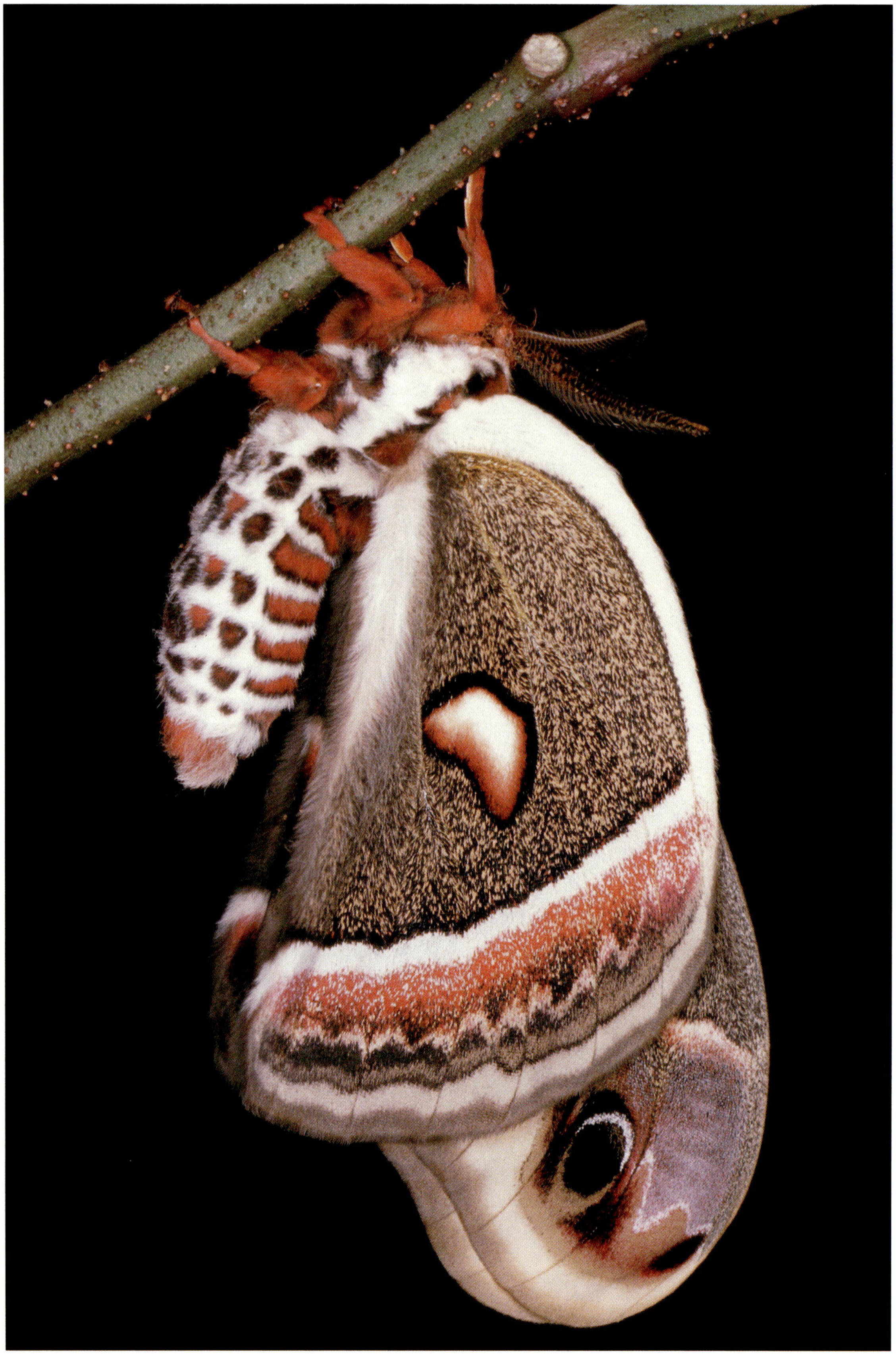

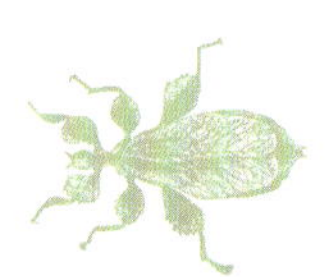

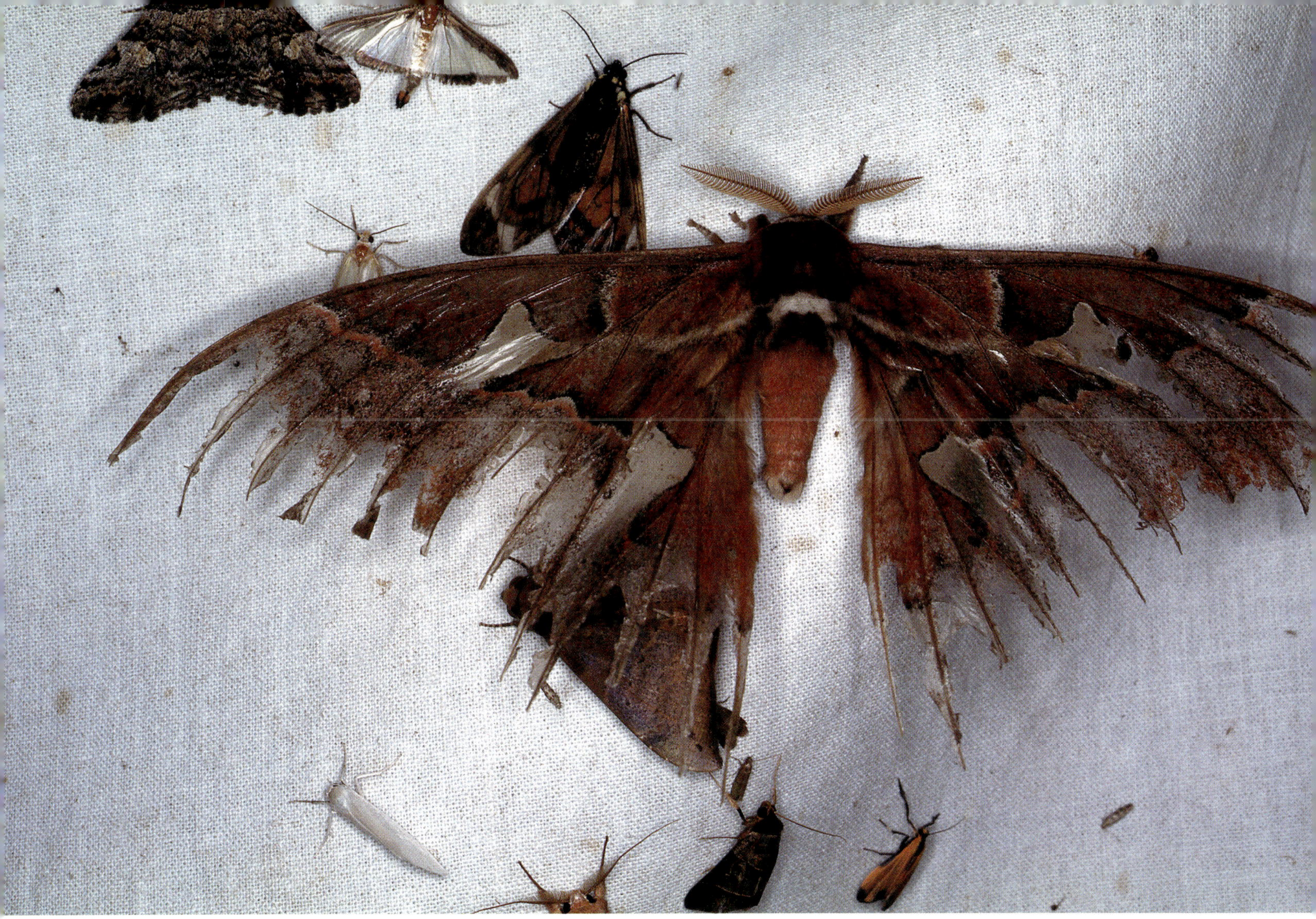

Sic transit gloria mundi (Thus passes away the glory of this world). The glorious emergence of the adult insect, at the peak of its powers, is followed by its rapid decline. This occurs over a period of days, weeks, or, in rare cases, a few months.

Above: A saturniid moth from tropical America *(Rothschildia morana)* is barely identifiable here.

Opposite page: A silkmoth from eastern North America *(Hyalophora cecropia,* family Saturniidae), newly emerged.

Thanks to chitin, nature exhibits an endless array of structures, finishes, and forms, all masterpieces of design. The strength and relative weight of an insect's protective outer covering (or integument) allow it to meet its specific requirements while also providing it with sufficient mobility in the air, in the water, or on the ground. Without these qualities, insects would never have made it past the drawing-board stage.

Some primitive insects, such as the silverfish still found in our homes today, have undergone relatively minor changes in comparison with the "modern" insects that began to appear more than one hundred million years ago. Nonetheless, the thin integument of silverfishes represents a considerable improvement over the simple methods used by soft invertebrates such as slugs, which cover themselves in a layer of mucus to conserve their essential fluids. "How can evaporation from the body surface be prevented?" asks Jacques Brosse. "Insects are bold inventors and skilled chemists:

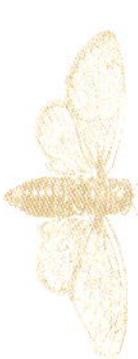

using their own sugars they synthesize chitin, which is nothing more than dried mucus" *(Encyclopédie essentielle).*

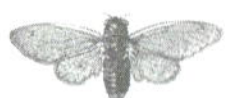

Before discussing the morphology of the adult insect, let us first consider the transformations that occur during metamorphosis, especially complete metamorphosis. This process occurs when an insect progresses from the egg to the larval (or caterpillar) stage and from there to the chrysalis or pupa, from which the mature adult insect, or imago, emerges.

The transformation of a caterpillar into a butterfly over a period of several weeks is in some ways even more impressive than the evolution of the archaeopteryx into the modern pheasant, which required several million years.

Strangely enough, larval stages may be of very long duration, ranging from several months to many years (seventeen years in the case of one species of cicada), while adult life spans can be very short, perhaps a few days or several weeks, depending on the species.

It is disheartening to think that all this effort and preparation can vanish so quickly in the final stage of life: the wonderful adult insect emerges, mates, and then crumbles to bits and pieces. Such is the fate of many insects, especially large saturniid moths and others with large wings that are easily torn during flight. Although this disproportion between life stages may seem absurd to us, it does not exist for the insect itself, which, of course, is equally alive and present at each stage of its development.[5]

Eating is the sole function of insect larvae, which continuously absorb nutrients that are necessary for their development into adulthood. In the final stage, the adult insect either continues to eat as it did during its larval stage, or eats very little; in some cases, it does not eat at all. Some adult insects possess completely different mouth parts than they did during their larval stage, and thus have very different diets. For instance, an adult butterfly typically uses its unfurled proboscis to gather nectar and other fluids, whereas it chewed and shredded leaves when it was a caterpillar. Adult

---

5. For the monarch butterfly *(Danaus plexippus)* these proportions are reversed; after a larval stage of about a month followed by fifteen quiescent days in the form of a chrysalis, the adult then enjoys a winged existence of a year.

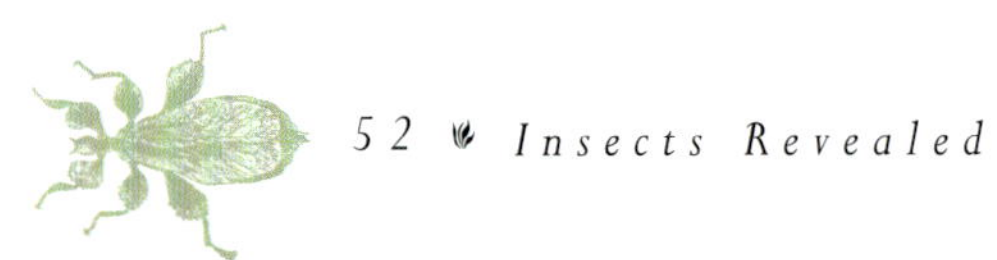

insects that refrain from feeding are less common; saturniid moths, which are in this category, have mouth parts that, for all practical purposes, are atrophied.

A larva's integument develops in such a way that available resources are used economically; its strength is no greater than that required by a sedentary organism in a generally protective environment. But the provisional nature of the integument is the most important reason for optimizing resources during this period. These eating machines grow rapidly, and their outer coverings are not extendable. As a result, larvae must molt periodically; they invest only those resources necessary for the production of this temporary skin and for ensuring the progression from one molt or developmental stage to the next. This process is repeated four to six times for most insect larvae and even more often among the larvae of primitive insects.

Amazing things happen during the development of the chrysalis: superfluous ambulatory and masticatory systems fade away and are either reconstituted in different forms or disappear completely. Other features appear as if by magic: wings, antennae, stingers, reproductive organs, even auditory mechanisms.

During the sleep stage of the chrysalis, for example, a butterfly retains only the first three pairs of legs of the caterpillar; the four pairs of false abdominal legs and the pair of anal legs disappear during the upheavals of metamorphosis. However, in the case of legless larvae, the three pairs of legs required by the adult organism must materialize. Thousands of transformations, either by addition or subtraction, occur among other types of insects. No matter what the

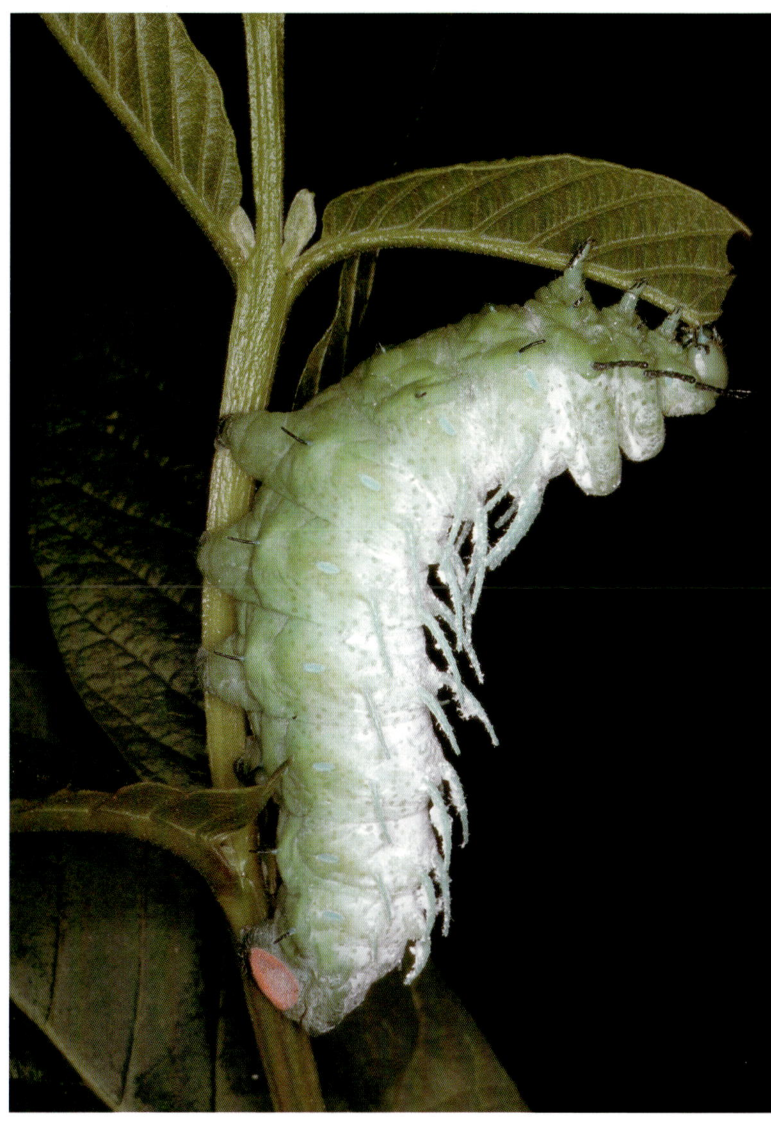

Of the caterpillar's numerous legs, only the three pairs from the anterior (or thoracic) region are retained by the adult.

Above: Caterpillar form of the atlas moth *(Attacus atlas),* a large nocturnal saturniid from Southeast Asia.

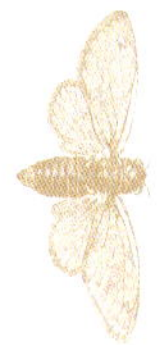

method, these transformations are always more drastic in those insects that undergo complete metamorphosis than in those that undergo simple metamorphosis. In the latter process, the progression from egg to adult takes place in stages leading to the emergence of the wings and reproductive organs, which constitute the only difference, aside from an increase in size, between adults and nymph. In the early developmental stages, for example, grasshoppers, cockroaches, mantises, and walking sticks closely resemble their adult counterparts. Dragonflies and other insects that have an aquatic nymphal stage undergo more profound changes since they eventually move from the water to the air, and must thus develop a completely different respiratory system.

In complete metamorphosis, butterflies or beetles must reconstitute themselves virtually from top to bottom. However, the beginnings of their adult body segmentation may be seen in the worm-like larva or caterpillar. During the final transformation, division of the body occurs, resulting in three separate sections: the head, the thorax, and the abdomen.

Lepidoptera
Butterflies or moths: Who's who?

The order Lepidoptera can be divided into two subgroups, the butterflies and the moths; in popular usage, these terms correspond to whether the insects are active during the day (diurnal) or the night (nocturnal).

All butterflies are indeed diurnal, apart from certain varieties that prefer dusk, such as the brassolids and the amathusids. Not all moths, however, are nocturnal. The family Uranidae, for example, includes as many diurnal as nocturnal species. The diurnal species have dazzling iridescent coloration, while those that are nocturnal are either white or dark colored.

If taxonomic categories are to be reliable and consistent, they evidently cannot hinge on such variable traits. The two groups have thus traditionally been differentiated by a more reliable trait, their antennal structure. The antennae of the butterflies are shaped like fine stems that end in a bulge or club-like structure; this group is known as the Rhopalocera. The antennae of the moths are so variable that these species are called the Heterocera, meaning different types of antennae.

In English, the term "butterfly" designates the Rhopalocera, whereas "moth" designates the Heterocera. Other languages also have popular terms that clarify this distinction. For example, in Spanish the corresponding terms for "butterfly" and "moth" are "mariposa" and "polilla"; in Portuguese, they are "borboleta" and "mariposa," although the latter has a slightly different meaning than it does in Spanish. French, however, lacks a popular term serving to differentiate between moths and butterflies.

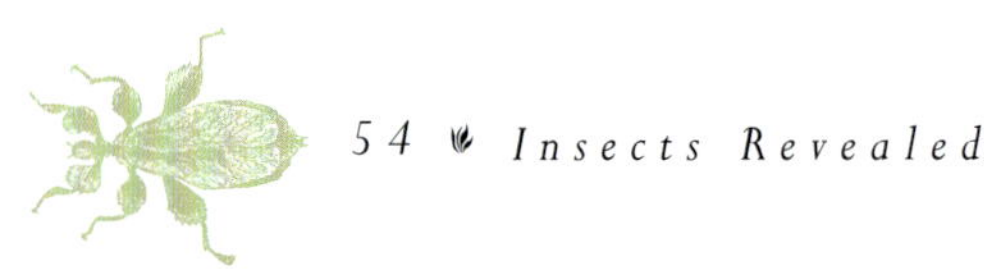

THE HEAD

The head comprises one whole segment, and features one pair of compound eyes, one pair of antennae, and some form of oral apparatus for crushing, sucking, or piercing. Some insects have simple eyes, whose function is still uncertain.

THE THORAX

The thorax is subdivided into three sections that are not always easy to tell apart: the prothorax, the mesothorax, and the metathorax. One pair of legs is attached to each of these subsections. In insects with two pairs of wings, the anterior wings, or forewings, are attached to the mesothorax and the posterior wings to the metathorax. The thorax is thus the locomotive center and contains a strong muscle system that controls both walking and flying.

THE ABDOMEN

The abdomen is made up of a maximum of eleven segments, which are attached to each other by means of a flexible membrane that allows for great mobility in all directions. A pair of respiratory orifices, or spiracles, is located on the side of each segment. The very last segments, which are sometimes fused together, are the site of the insect's excretory and genital organs, ovipositor (the egg-laying organ), stinger, and other appendages, depending on the species.

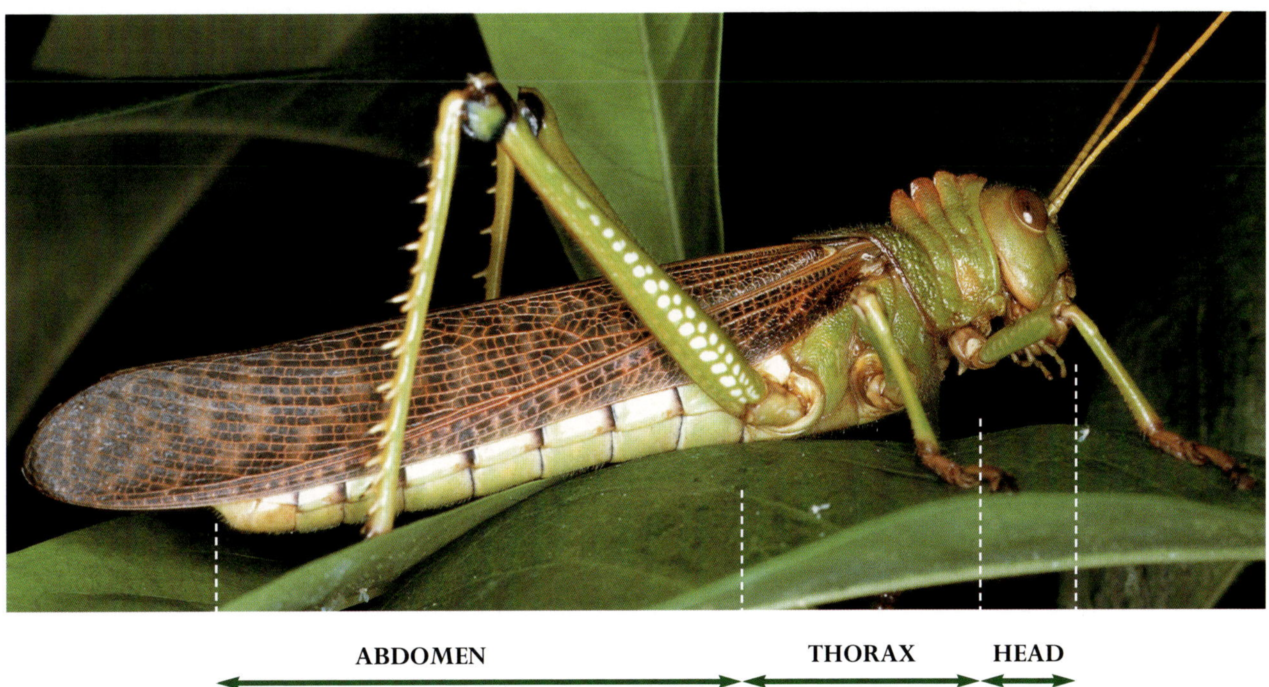

The three body sections of an insect: abdomen, thorax, and head.

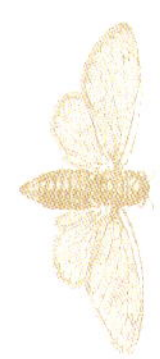

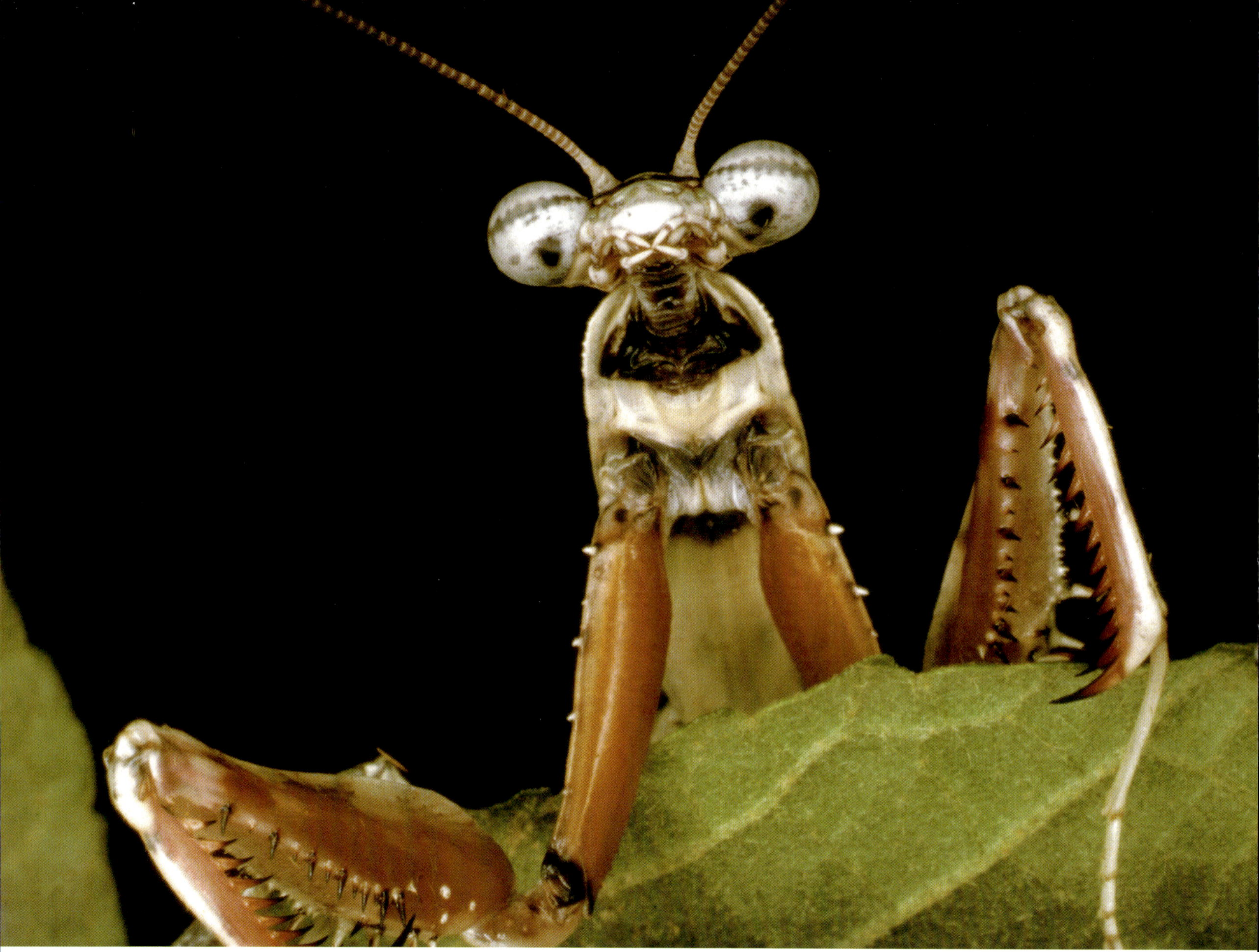

THE LEGS

During the evolution of the very first insects, legs appeared several million years before wings. This accounts for the huge gulf between the primitive wingless insects and hexapods (apterygots) and the winged pterygots.[6]

I have mentioned the apterygots only in passing, but before consigning them to their distant past, I should point out that during the time between the development of legs and that of wings, evolutionary processes were quite uninventive. The apterygots' six legs had a single function, locomotion. It was not until millions of years later, after thousands of new types of insects had appeared, that certain legs, most notably the anterior and posterior legs, began to evolve to perform functions other than simple locomotion. Legs designed for burrowing, jumping, capturing prey, and swimming illustrate the marvelous adaptations that accompanied the growing diversity of living things.

6. From the Greek *pteron,* wing. This root appears frequently in entomological vocabulary, especially in the names of the orders (for example, Lepidoptera and Diptera).

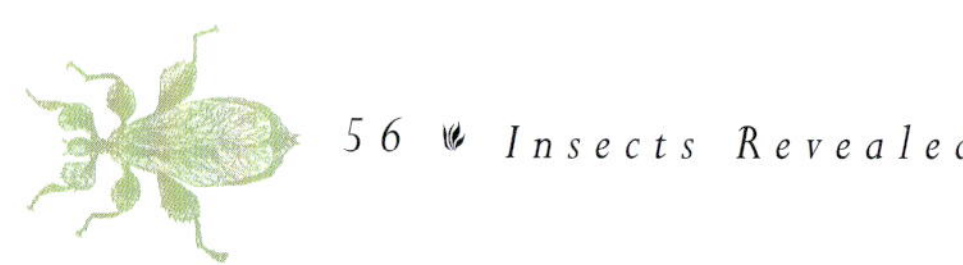

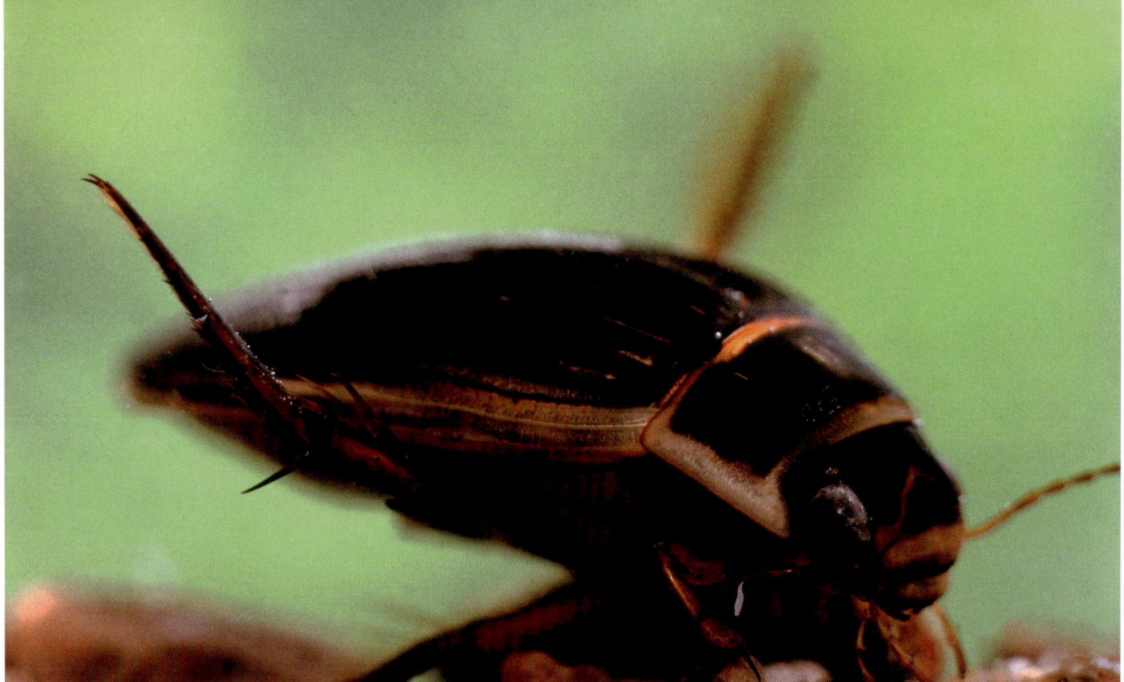

The legs' primary function is locomotion, but many others have developed during the course of evolution: mantises' legs are used to capture prey, aquatic insects have ciliated legs, and so on.

Above: An aquatic beetle from eastern North America (*Dytiscus harrisi,* family Dytiscidae).

Opposite page: Mantis from the Dominican Republic.

Insects' six legs are attached in pairs to the three subdivisions of the thorax. At its base, each leg includes two inconspicuous segments, the coxa and the trochanter; the femur, the tibia, and the tarsus comprise the more recognizable portion. The tarsus is divided into several smaller components, the last of which extends into a pair of tarsal claws. The variations on this theme are endless: the anterior legs may become disproportionately longer than the middle or posterior legs, as is the case for the harlequin beetle of tropical America *(Acrocinus longimanus),* certain weevils indigenous to Southeast Asia, or the scarab beetles (subfamily Euchirinae) from the same region. Other scarab beetles (subfamily Rutelinae) from Central and South America reverse these proportions and have extremely robust posterior legs, most notably in the femur. In all these cases, only males display such exaggerated proportions.

According to the needs of different insect species, specialized accessories may also be located on parts of the first or last pair of legs. These structures, which do not fundamentally alter the form of the legs, have numerous functions: cleaning the antennae, producing or perceiving sounds, adhering to smooth surfaces, and so on. For example, butterflies' taste sensors are located in the tarsal region of their front legs, and grasshoppers of the family Tettigoniidae hear by means of eardrum-like tympana found on the tibiae of their front legs. (The sense of hearing is very rare among insects; in addition to grasshoppers, those with this ability include other members of the order Orthoptera that make shrill trilling sounds, along with cicadas and other insects with sonic or infrasonic receptor organs.)

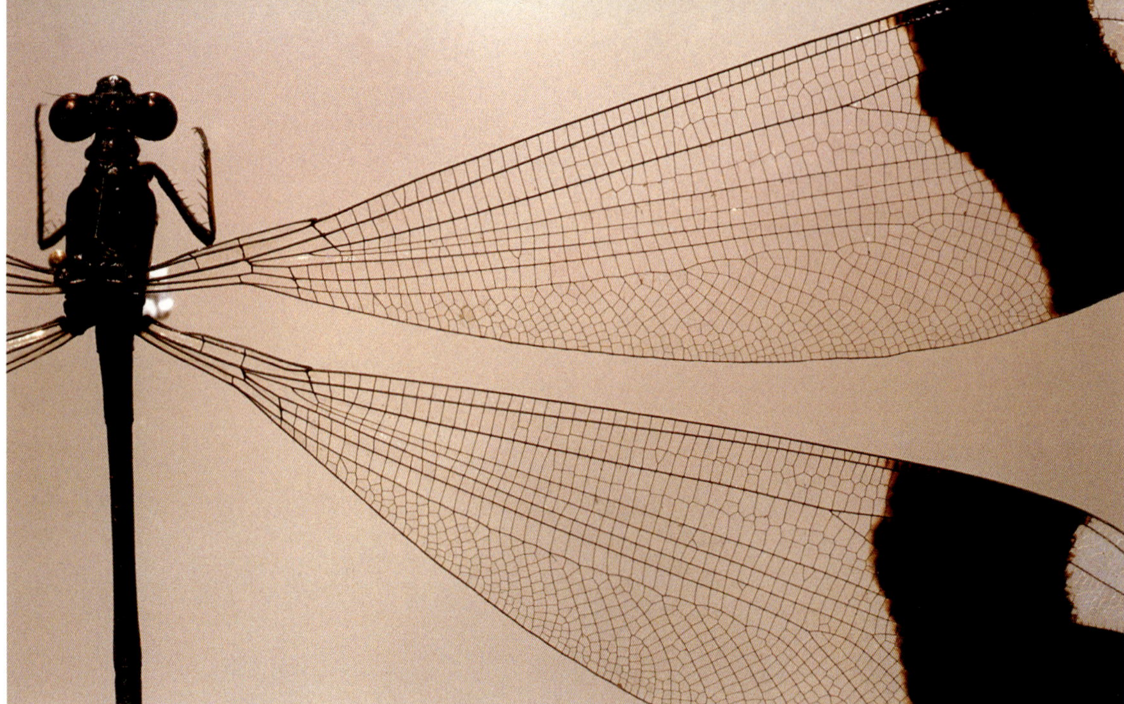

The process of evolution has usually modified the insect leg via the process of addition. However, among the many brush-footed nymphalid butterflies, a subtractive process has occurred: the locomotive capacity of the front legs has been lost, and they now serve as feelers. The nymphalids manage as well with four legs as other varieties do with six.

THE WINGS

Although the larvae of most insects already possess the adults' six legs, those that undergo complete metamorphosis show no visible signs of what will later constitute the wing structure. The wings emerge during the pupal phase and are clearly outlined on the chrysalis or pupa. Among insects undergoing simple metamorphosis, the wings emerge during the final molts.

The wings of primitive insects such as dragonflies are characterized by finely veined structures resembling netting or lacework, with long and narrow structural

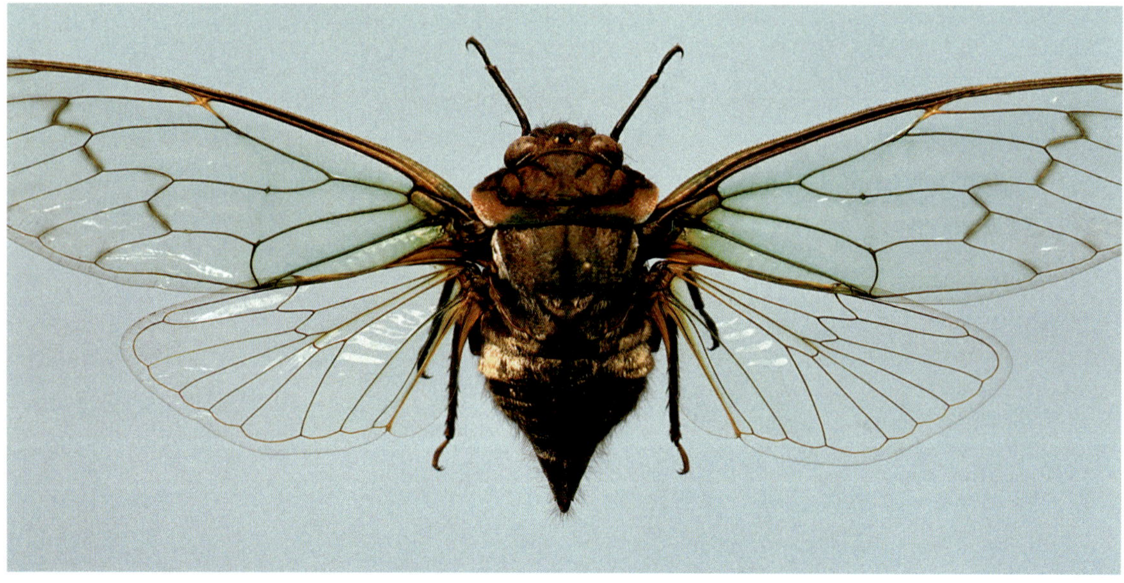

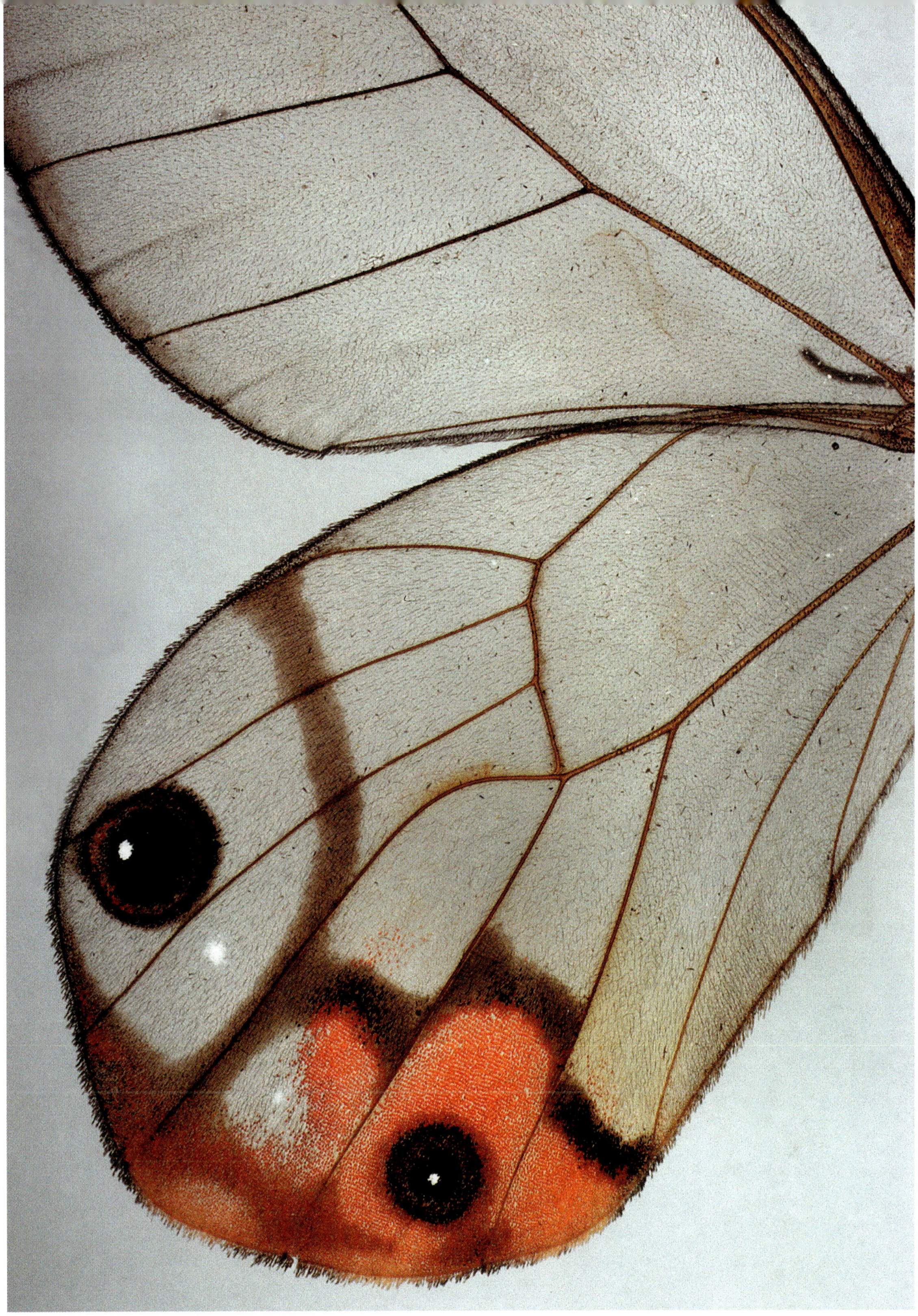

Wings have gradually evolved into more simplified structures.
Above: Wing of a butterfly (genus *Haetera,* family Satyridae) from tropical America.
Opposite page, top: Wings of a large damselfly *(Megaloprepus coerulatus)* indigenous to tropical America.
Opposite page, bottom: Wings of a cicada from New Guinea.

supports. This nerve-like network was reinforced when other types of wings began to emerge: the veins became simplified and were organized into reinforced primary veins, from which the secondary veins branched off; this enabled the wings to become wider and increased their structural qualities. The wings of butterflies and cicadas are excellent examples of this phenomenon.

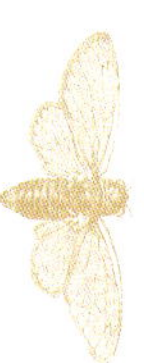

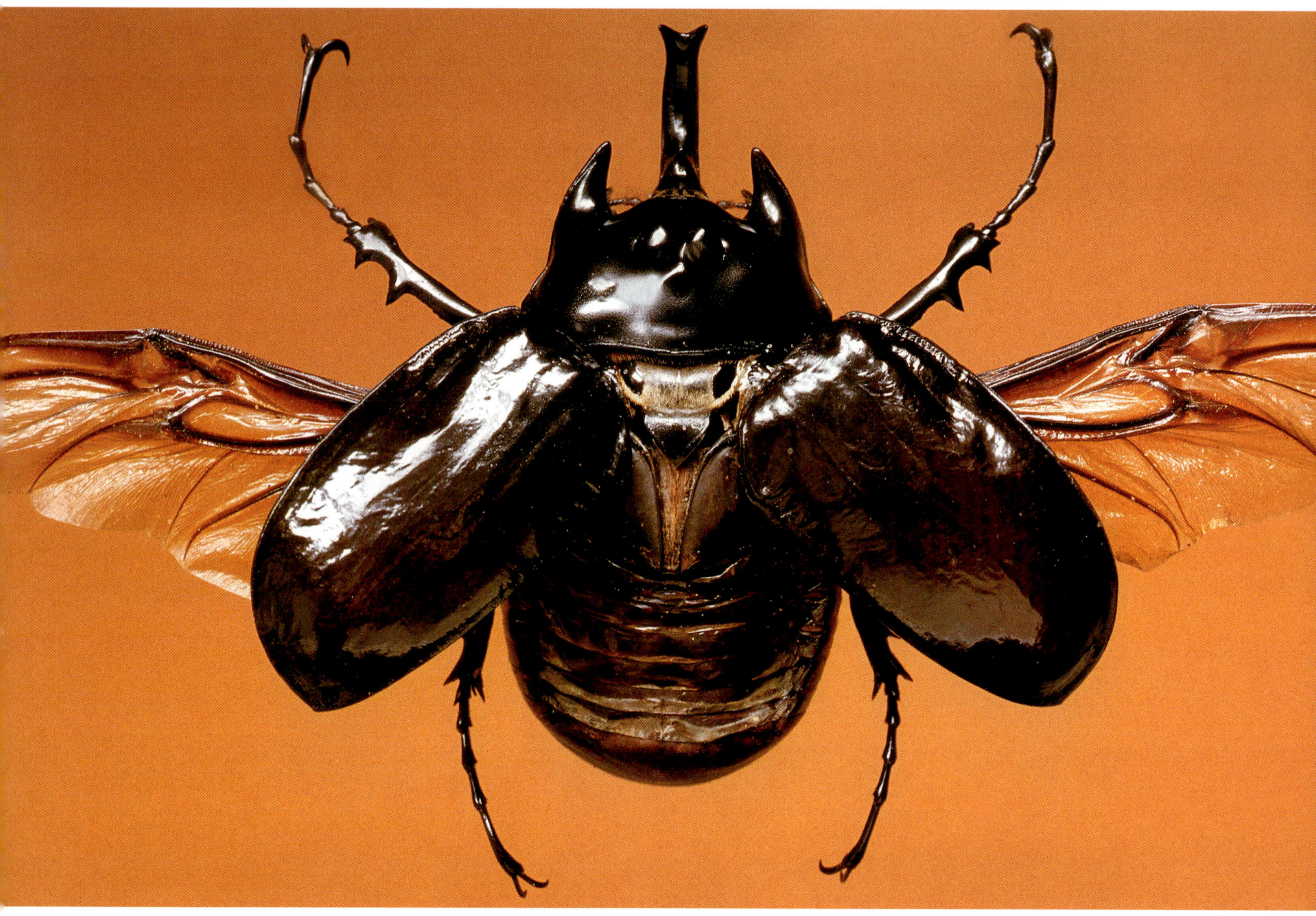

Among the beetles, the forewings serve as a protective covering for the posterior wings, which are used for flying. Shown here: a scarab beetle from tropical America (*Megasoma acteon janus*, subfamily Dynastinae). Wingspan: 23 cm. (9 in.).

Beetles' wings underwent more radical modifications: the forewings were transformed into hardened sheaths, or elytra, that protect the posterior wings, which are used for flight. Once more, evolution has worked its wonders. Beetles are often robust and massive, and could be supported only by strong, large wings. But how could such large wings be hidden beneath elytra that only half cover them? The mystery is explained by the articulation of the membranous and elbowed wing structure, which folds back on itself and disappears instantly beneath the protective sheath. This phenomenon is most pronounced among the rove beetles of the family Staphylinidae.

In another extraordinary transformation—similar to the process that transformed the anterior legs of the nymphalids into feelers—the posterior wings of the Diptera (flies) have shrunk to form a sort of gyroscope, a high-frequency vibrating mechanism that serves to stabilize the insects during flight.

The "inventions" that contributed to the development of insects' wings transcend those that turned insects' legs into the multipurpose tools that must have inspired inventors of the past. Engineers and designers could still find much to admire in both types of structures; they might even conclude that all of their ideas had already been anticipated!

In keeping with their name, the pterygots should all be winged insects. Nevertheless, even here we come across confusing exceptions: if wings are standard, why are the females of some species wingless?

Insects' exoskeletons set limits on their body mass and size. Some 250 to 300 million years ago, during the Carboniferous era, one variety of dragonfly *(Meganeura monyi)* had a

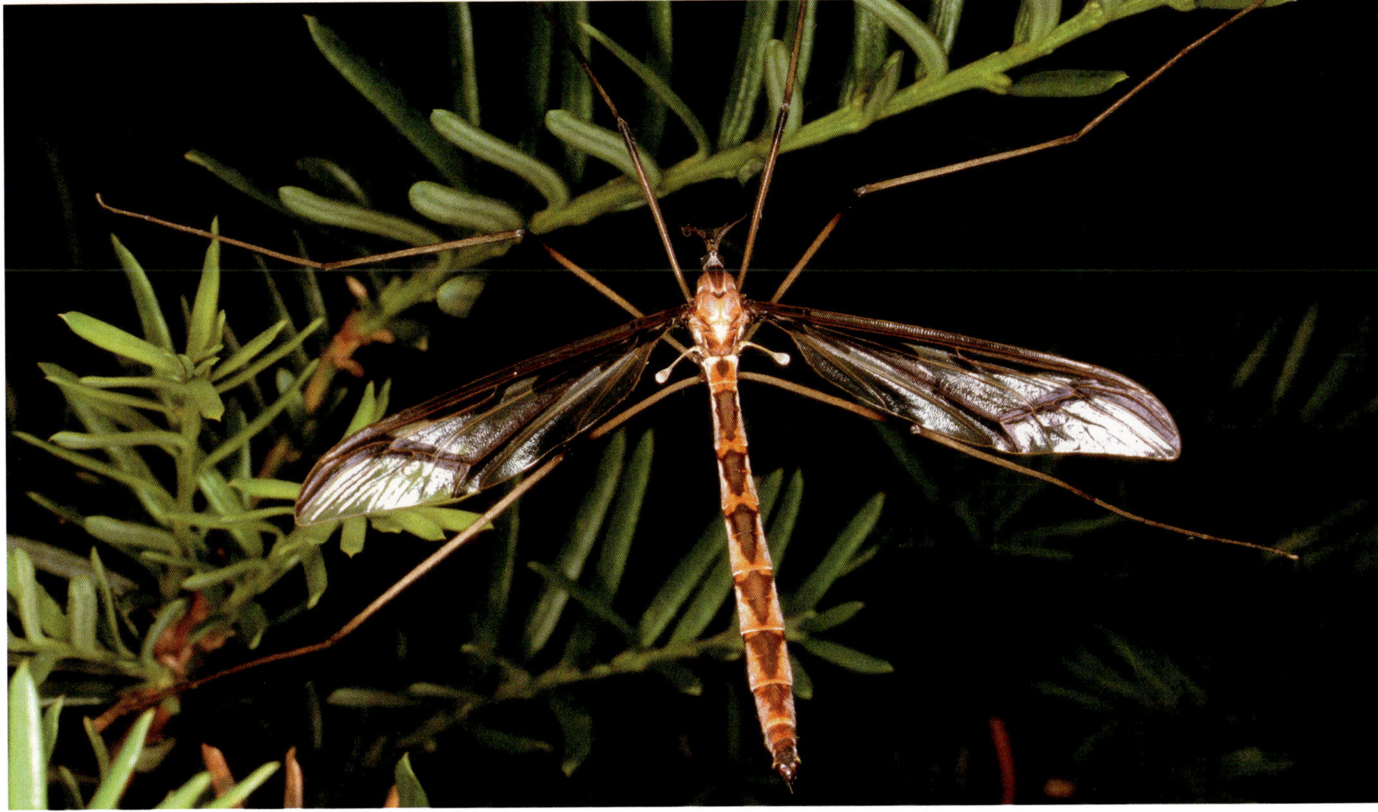

By an extraordinary process of transformation, the posterior wings of the Diptera have evolved a unique function in the insect world: shaped like small dumbbells, they act as a high-frequency vibratory apparatus that serves to stabilize flight. In this large cranefly (family Tipulidae) from eastern North America, this apparatus is located at the base of the thorax.

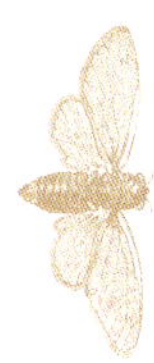

70-cm. (27.5-in.) wingspan. Fossil remains offer clear evidence of its enormous size; scientists know of no other insect that has attained such dimensions. It is now extinct, together with most of the other Carboniferous-era species. Could its disappearance be related to its size? We shall never know the answer because the smaller species that coexisted with *Meganeura monyi* are also extinct. Whatever the case may be, the march of evolution has served to scale back the size of today's giants, which have more modest proportions and perhaps greater viability. These include the goliath beetles of equatorial Africa and the gigantic Dynastinae and Prioninae of tropical America, together with the phasmids of Southeast Asia and the large saturniid moths from the same region.

THE MOUTH PARTS

Across the various insect orders, mouth parts are so complex and variable that I limit myself here to a simplified overview.

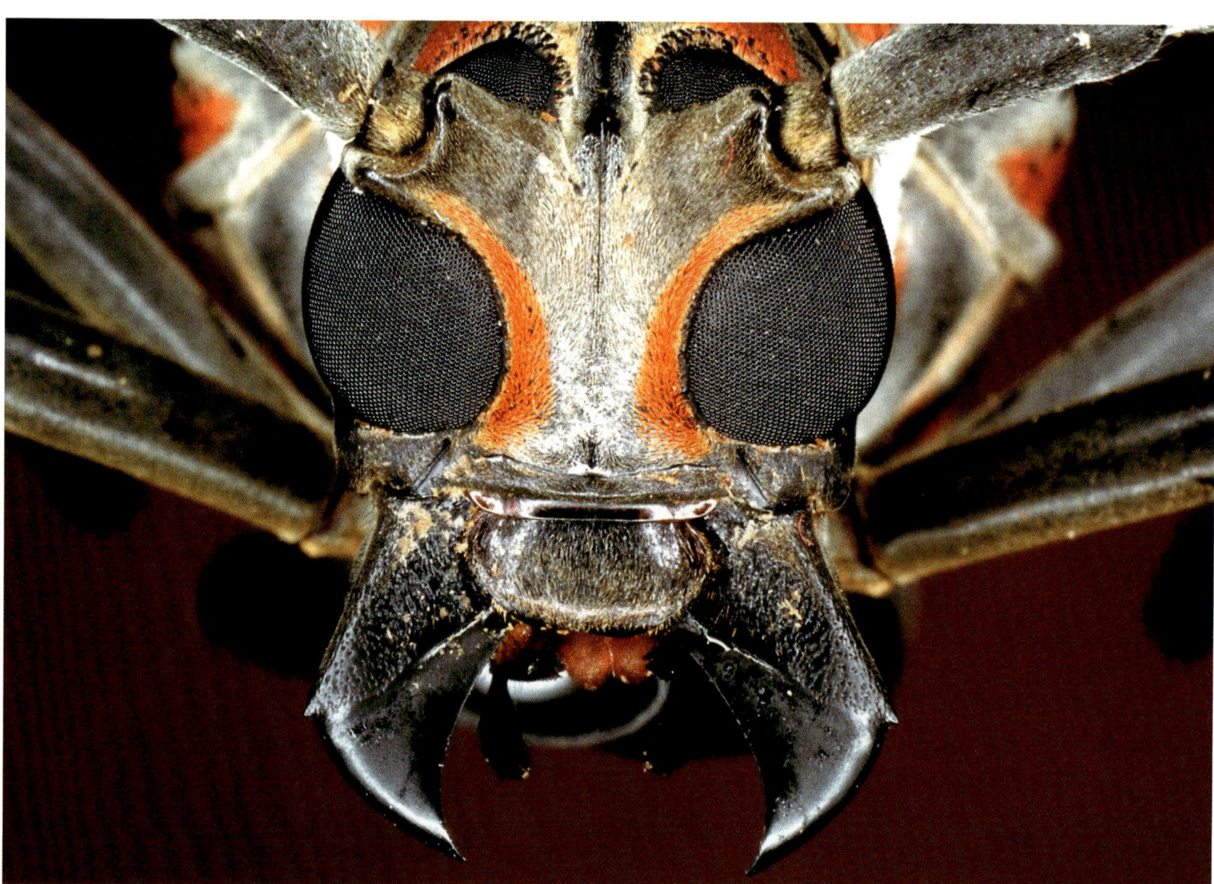

The basic components of the mouth parts are used to form many different structures. Shown here: the head of a magnificent cerambycid beetle (*Rosenbergia* sp.) from Papua New Guinea. Note the amount of space occupied by the compound eyes. As in many insects, the broad visual field compensates for the fixed position of the eye in a head that is generally immobile.

The same fundamental components constitute (and reconstitute) the structures used for chewing (beetles, grasshoppers, dragonflies), sucking (butterflies), licking (bees), piercing and sucking (cicadas, fleas), biting and licking (flies), and so on. Such modifications and adaptations for specialized feeding purposes do not necessarily make use of all the basic components of the insects' oral apparatus; some may be atrophied or even completely absent. In a few groups, like the saturniid moths, the entire oral apparatus may be reduced to a vestigial form; the adult insect thus lives on nutrients absorbed during its larval stage and has a fairly short life span.

THE EYES

The compound eye, which had already evolved by the Carboniferous era, allowed winged insects to navigate and to capture airborne prey with new-found speed. These insects almost certainly had a highly developed sense of sight, and although the eye structure has undergone various subsequent improvements, it has not changed markedly. As in the case of other examples we have seen, the compound eye has become increasingly diversified over time.

The main characteristic of insects' eyes is that they are fixed. To compensate for this, their visual field is quite wide, especially among hunting insects such as dragonflies or tiger beetles, together with insects like flies, which flee at the slightest threat. These insects are hard to approach because their eyes extend around their head.

These protruding eyes are made up of cone-shaped units or facets known as ommatidia. Each ommatidium is a functional six-sided eye surrounded by several thousand others; certain large dragonflies may have as many as 28,000 of these units. Typically, a cluster of ommatidia forms the shape of a globe, although some species exhibit more unusual shapes, such as those in the family Cerambycidae (longhorn beetles), whose eyes wrap around the base of the antennae, extending from the front all the way to the top of the head.

Whether attacking or defending, the most active insects are highly sensitive to images that shift from one facet of their eyes to another, especially rapid-fire images. A butterfly will tolerate the presence of a photographer if he or she approaches slowly and cautiously, but will fly away as soon as the shutter release button is pressed.

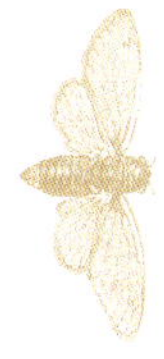

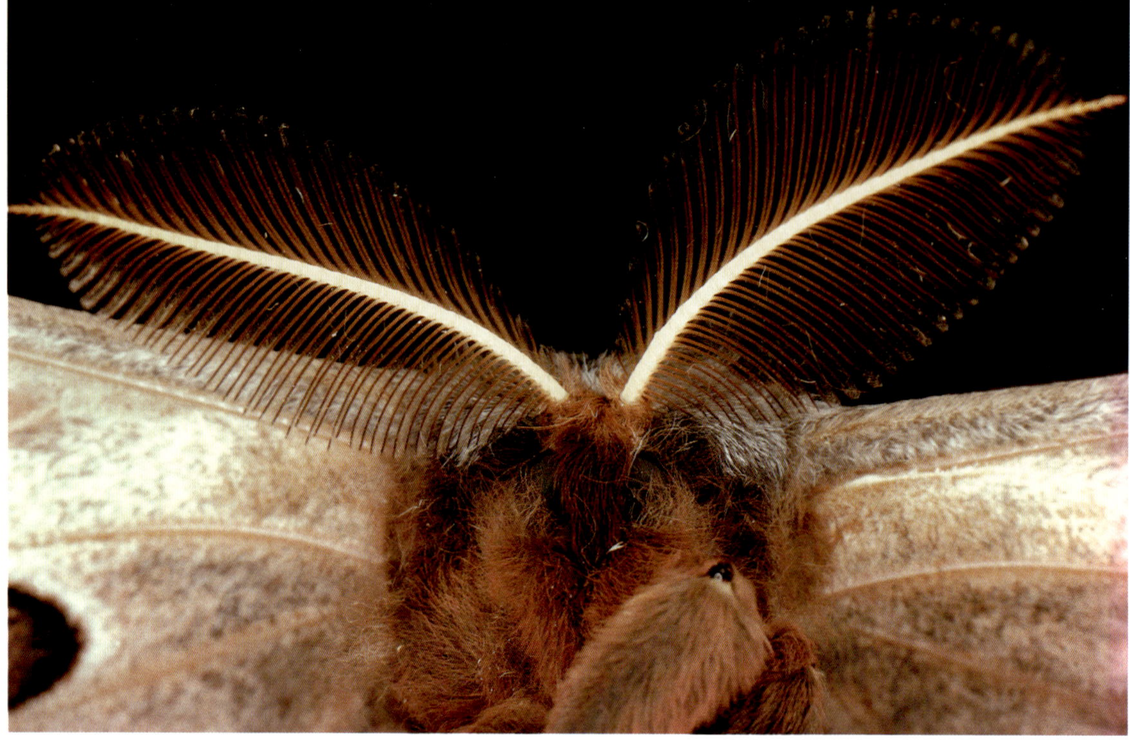

The form, structure, and length of insects' antennae vary greatly, but they all have the same function: to allow the organism to sense the world around it.

Above: The superb plumed antennae of a large male saturniid moth *(Antheraea polyphemus)* from eastern North America.

THE ANTENNAE

Antennae are virtually exclusive to the insect world, although some other arthropods, such as crustaceans and myriapods (centipedes and millipedes), have antennae as well.

Antennae play a key role in insects' olfactory and tactile perceptions throughout their life spans. As Walter Linsenmaier has observed, "Using its antennae to smell and probe at the same time, an insect is able to detect fantastically dilute odors. . . . In this way, an insect finds the proper plants for feeding or for depositing eggs, a parasitic female tracks down its animal host, and an insect of one sex seeks out the opposite sex" (*Insects of the World* [New York: McGraw-Hill, 1972], p. 22).

In addition to being extremely sensitive, antennae come in a wide variety of shapes and sizes, ranging from a pair of tiny strands (cicadas and dragonflies) to huge articulated whips more than 20 cm. (8 in.) long among certain longhorn beetles of New Guinea, or the superb feathery structures peculiar to male saturniid moths.

Some important taxonomical distinctions are based on antennal structure. For example, the two divisions of the Lepidoptera are differentiated by the shape of their antennae. Butterflies, which have threadlike antennae (cilia) ending in a slight bulge or club-shaped structure, are known as the Rhopalocera (from the Greek *keras,* horn, and *ropalon,* club), whereas moths are called the Heterocera because they have many different types of antennae (from the Greek *hetero-,* different).

INTERNAL ANATOMY

It may be surprising to learn that insects' respiratory systems are not connected with the mouth or nose, as we might naïvely assume. In fact, insects do not have noses, and their respiratory tracts are completely dissociated from their mouth parts. All of the abdominal segments and two segments of the thorax feature a pair of spiracles that transmit air to the cells by means of tube-like tracheae that branch out progressively to form tracheoles, the fine terminal branches of the respiratory tubes. (An insect's "blood," or hemolymph, does not circulate oxygen throughout its body by means of a vascular system, as is the case among the vertebrates.)

Air sacs located along this network activate respiration, either by normal bodily movements or by voluntary action.

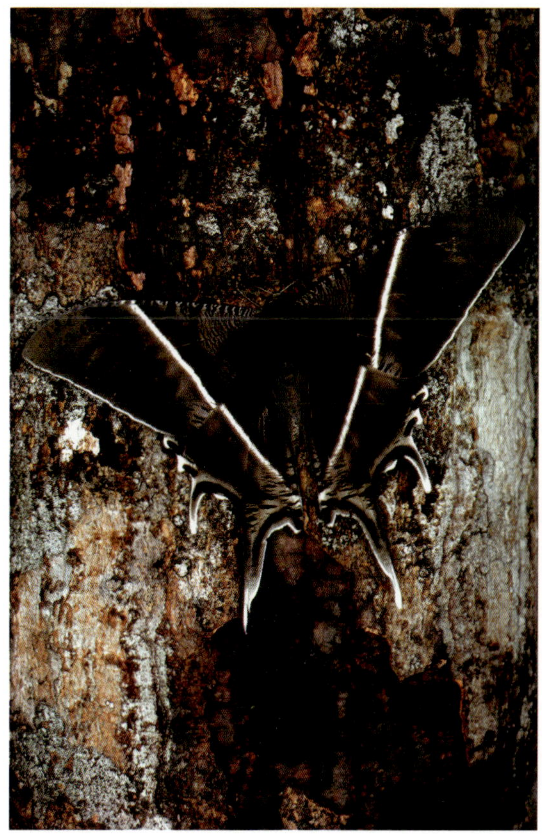

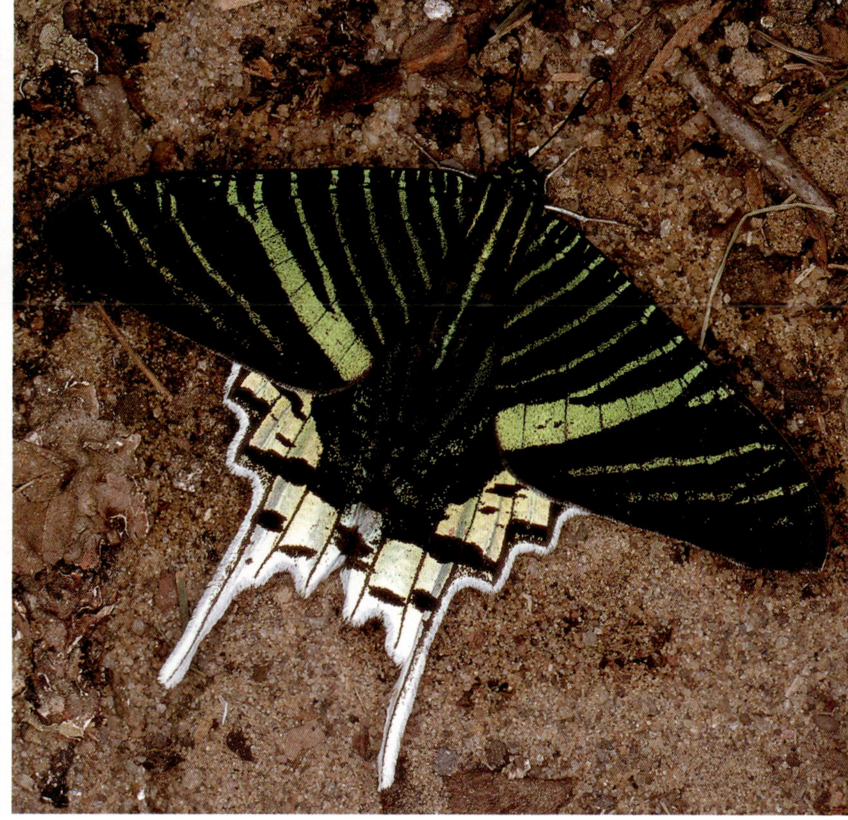

This giant swallow-tailed moth (*Lyssa zampa docile,* family Uranidae), which is widespread throughout Southeast Asia, is true to the vocation of most moths: it flies at night.

This day-flying moth (*Urania leilus,* family Uranidae) is found throughout the Amazon basin. If it flew by night, like the swallow-tailed moth, what purpose would its iridescent coloration serve?

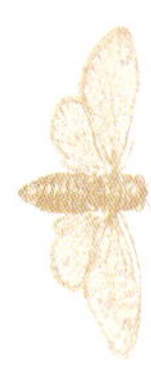

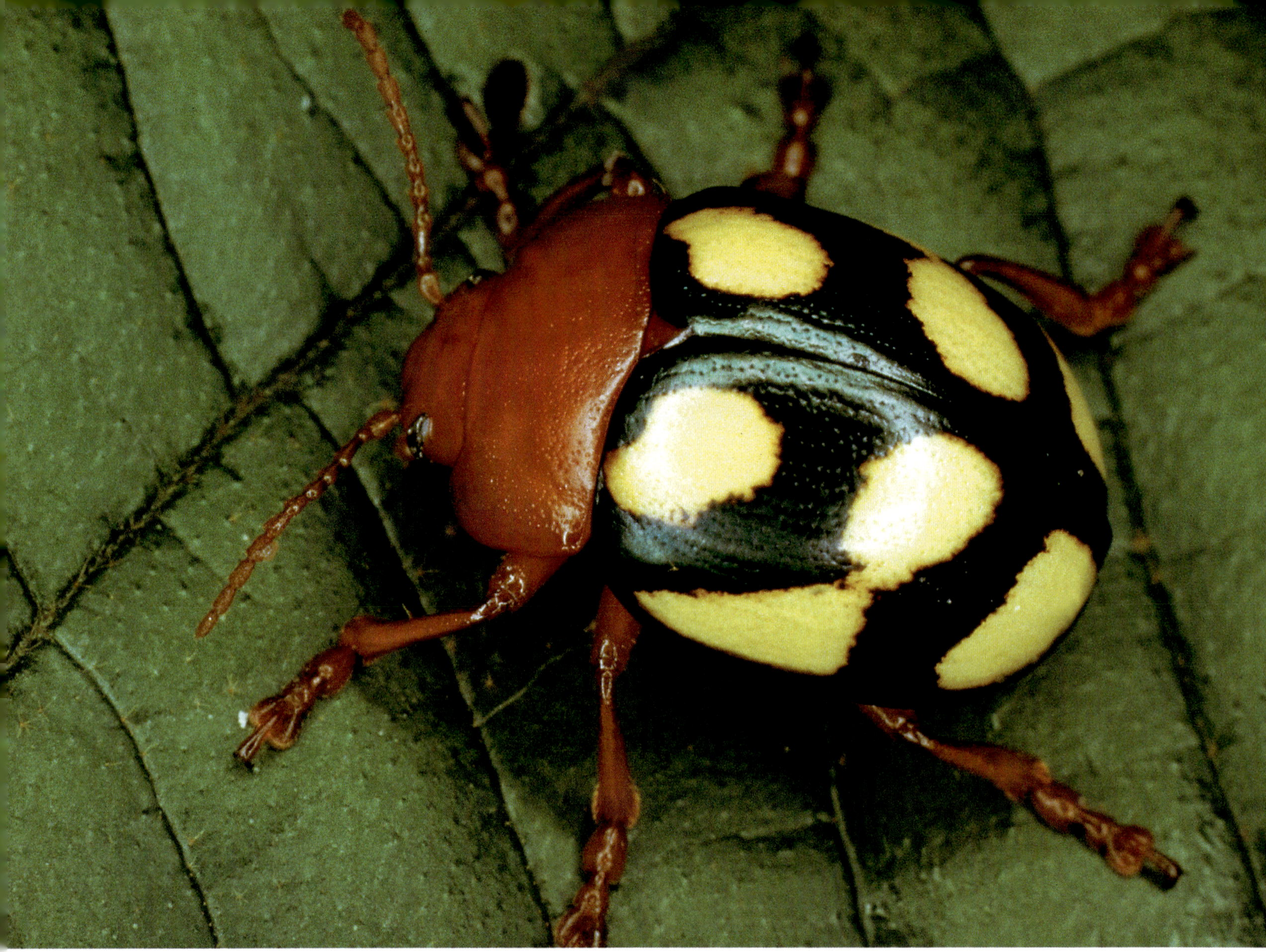

When attacked, many insects, such as this tropical American chrysomelid beetle, secrete drops of their toxic hemolymph through their joints.

Hemolymph is a clear, almost colorless fluid that circulates freely within the exoskeleton. It is pumped by the heart, an elongated vessel extending along the dorsal region, and distributes only assimilated nutrients to the cells. Certain insects use their hemolymphatic pressure for a number of purposes: a newly emerged butterfly quickly spreads its wings by increasing the hemolymphatic pressure through the wings' veins, and insects whose hemolymph is toxic squeeze several drops out through their joints when attacked.

We may be surprised to learn that an insect's circulatory apparatus is located in its back, while the central nervous system begins at the head and continues along the underside of the body. At the base of the ganglionic chain (a collection of nerve cells), the brain receives and processes the stimuli transmitted by the sensory organs, and then issues commands that determine action and behavior. Insects are born with predetermined instincts; their life cycles are too short for them to learn by individual

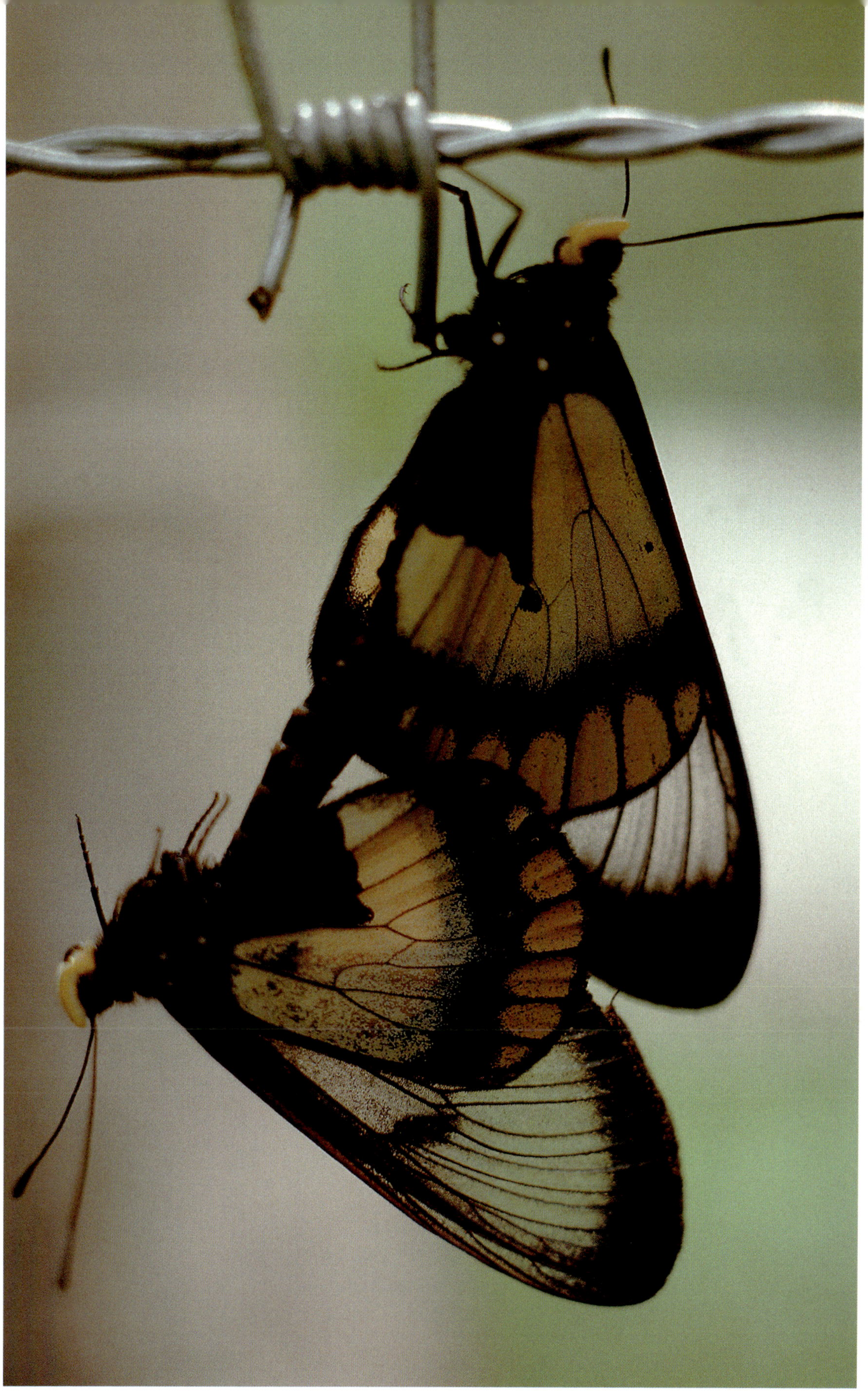

Butterflies (family Acraeidae) mating in Papua New Guinea.

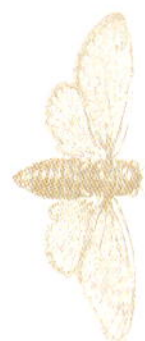

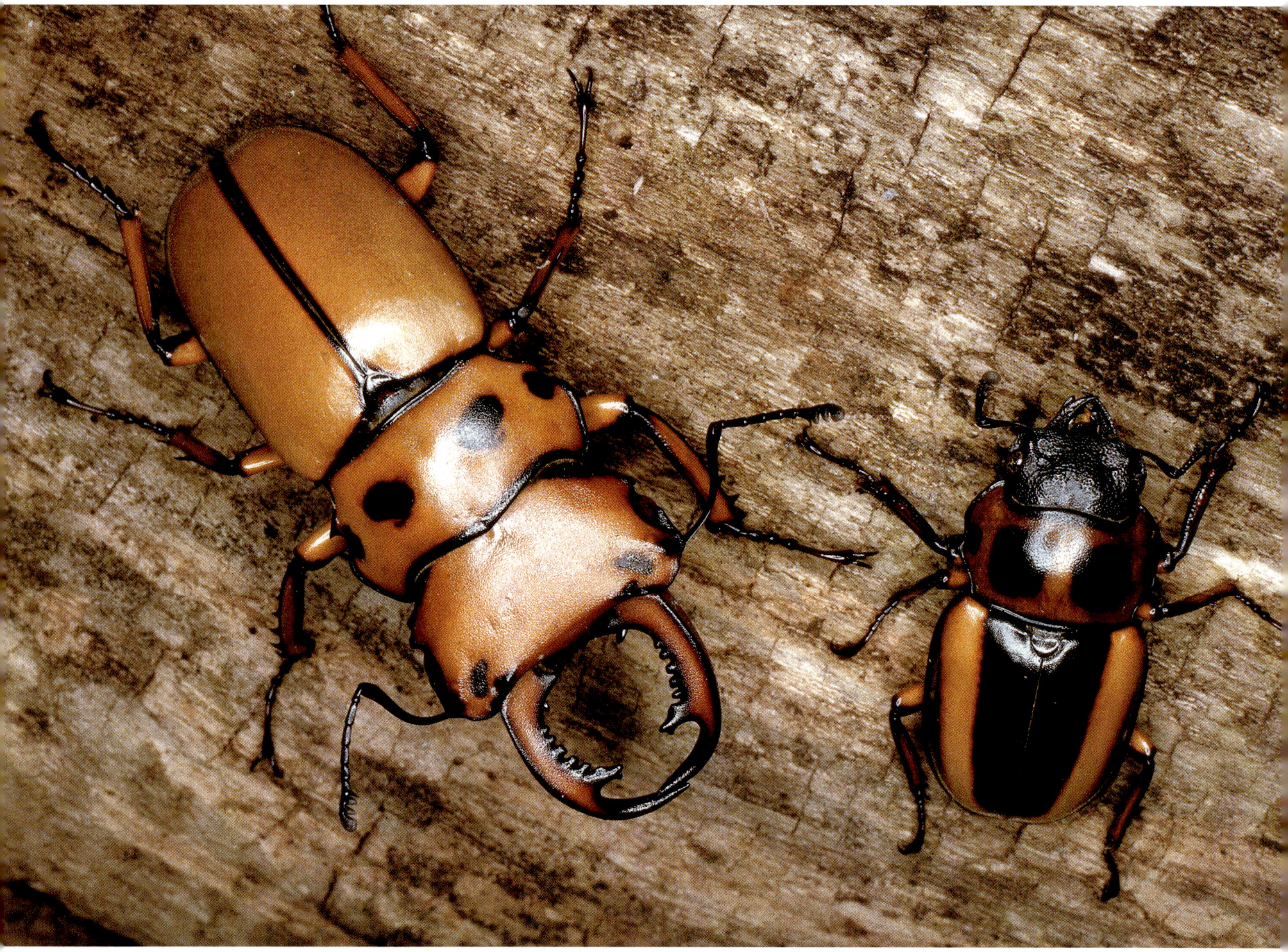

Sex differences are often evident in differences in length of mandibles, antennae, legs, or other appendages; the plumed antennae of the saturniid moths exemplify this phenomenon. In some cases, size and color differences accompany the primary attributes.

Above: Male and female beetle of the species *Homoderus mellyi* (family Lucanidae) from equatorial Africa.

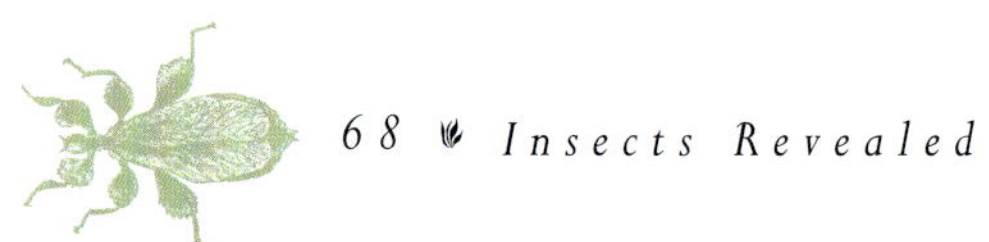

experience. If insects' genetic programming allows for any measure of freedom at all, it is an extremely limited one.

Like almost all animals, insects mate to propagate themselves. However, rare cases of parthenogenesis occur, especially among certain species of phasmids among which nonfertilized females lay eggs that produce only other females; in fact, males are either unknown or nonexistent. Among social insects such as bees and ants, fertilized eggs produce female larvae, and nonfertilized eggs become males.

Some insects exhibit extreme sexual dimorphism. The differences in form, color, size, antennal structure, or possession of wings are sometimes so pronounced that males and females were often thought to belong to different species.

Although insects' digestive apparatuses have few remarkable characteristics, the phenomenon of symbiosis is noteworthy. Symbiosis occurs as part of the digestive processes of certain species that consume "indigestible" substances such as wood, wool, or the horny parts of dead animals, which contain cellulose, lignin, and keratin, respectively. Aside from species that use enzymes to break down cellulose, other consumers of these substances rely on symbionts, microscopic bacteria or protozoa that live in the larva's "fermentation chamber" and facilitate digestion.

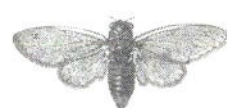

Since many details pertaining to insects' senses, functioning, and behavior lie outside the scope of this book, I must show restraint in describing these aspects of the real insect—which, as noted above, is much more fascinating than anything the human imagination could ever dream up.

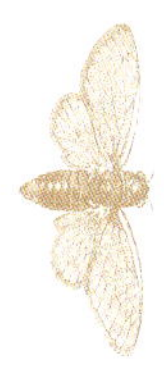

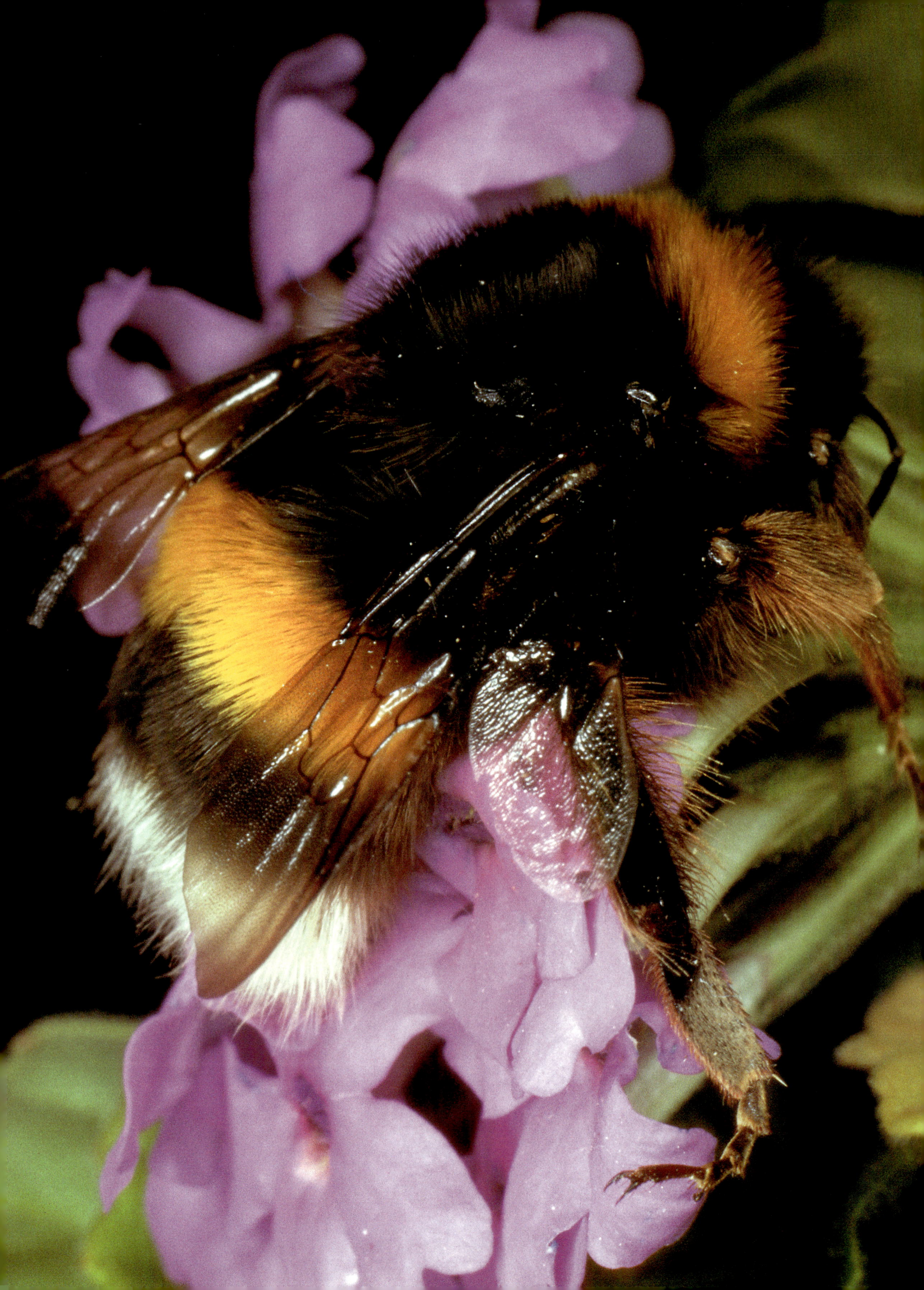

insects and temperature

Unlike the warm-blooded vertebrates, the arthropods (including insects) have no means of producing or regulating internal heat. However, it would be wrong to conclude that arthropods are literally "cold-blooded"; their body temperature, in fact, matches that of their environment.

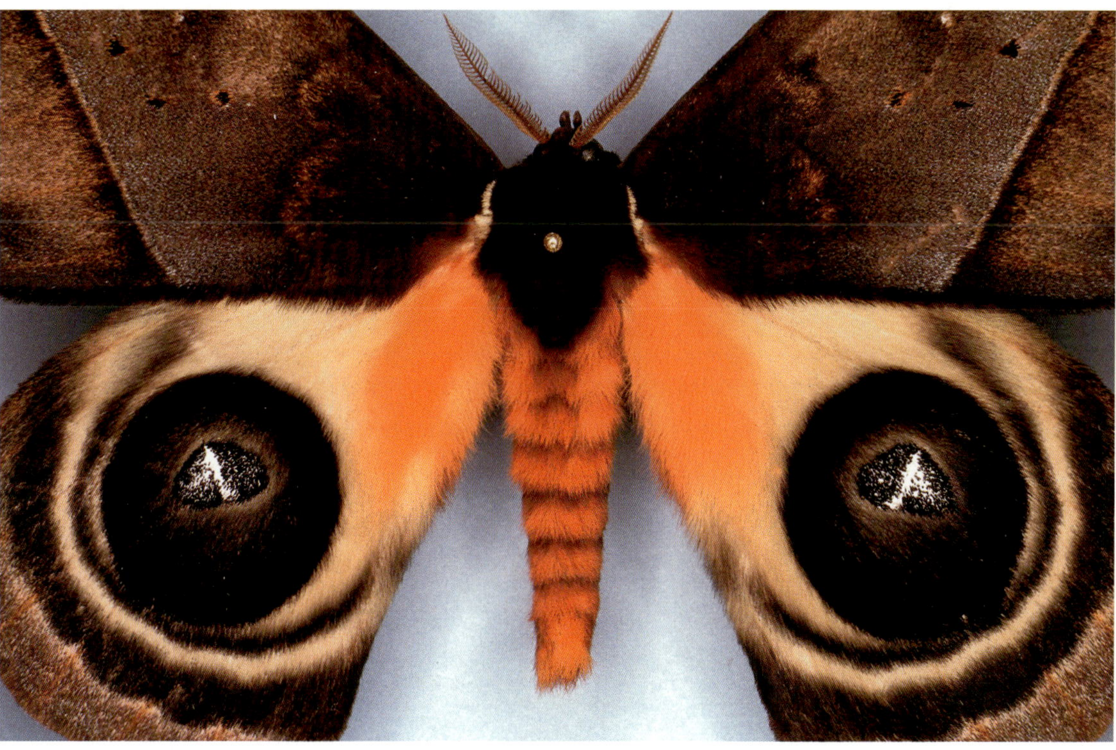

Hair on the body and wings of several insect species is more than merely decorative.

Above: A moth *(Automeris postalbida)* from tropical America.

Opposite page: A bumblebee *(Bombus terrestris)* from Europe.

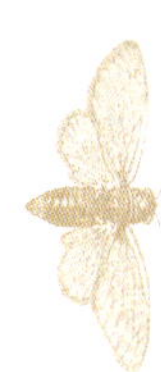

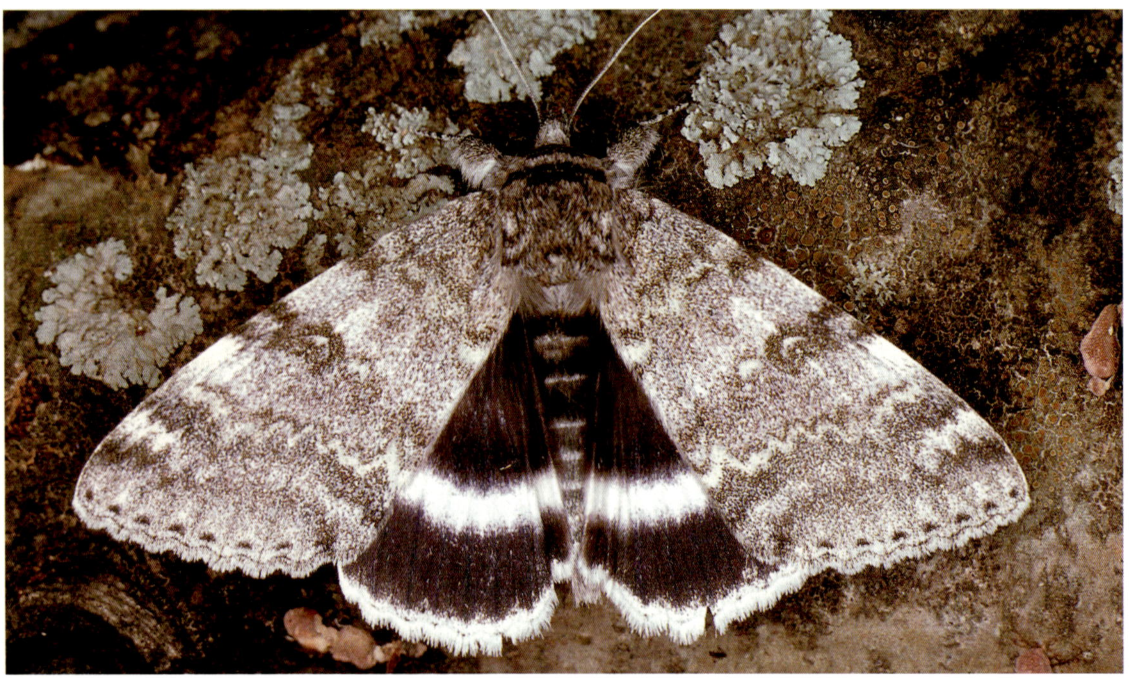

The white underwing moth *(Catocala relicta),* a noctuid from eastern North America. The *Catocala* from northern regions exhibit an astonishing tolerance to cold.

This physiological trait is extremely important, since it has influenced the dispersal of species throughout the world's climatic zones. Because an insect does not produce body heat, it does not enjoy the same degree of autonomy as do mammals and birds. When faced with shifts in temperature, insects must adapt, just as they have for millions of years; examples of such adaptations are as numerous as insects and climates themselves. The most straightforward response to hostile conditions, however, at least for species with adequate mobility, has often been simply to find a more hospitable habitat nearby.

The behavioral and physiological adaptations to climatic conditions that can be observed among insects today are so numerous and complex that I will summarize only their most essential characteristics.

As they emerge from the cool of the night, diurnal insects bathe in the warmth of the sun's rays; this quickly raises their internal temperature, and they can then go about their normal activities. We have all seen butterflies, dragonflies, or bumblebees at daybreak, covered in dew, lethargic and unable to fly. They are waiting for the air to warm up or for the sun to shine directly on them, accelerating the warming process. Birds greatly appreciate this time of day because insects that cannot flee are vulnerable indeed.

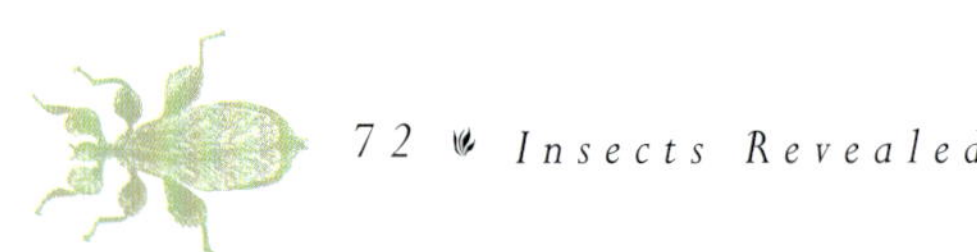

But what options are available to nocturnal insects, whose period of activity coincides with decreased temperatures? In general, they have adaptations that enable them to live at lower temperatures than can their diurnal counterparts. Moths normally have a thick insulating layer of fine soft hairs that enables them to maintain a temperature above that of the ambient environment. The British entomologist Vincent B. Wigglesworth remarked: "A bumblebee has a fur coat, and this can be quite important in keeping the thorax warm. If the hairs are clipped short a bumblebee cools down very much more quickly. The scales of moths have the same effect; they enclose a thin sheet of air that serves to insulate the body. A privet hawk moth *(Sphinx ligustri)* with its coat of scales intact had a body temperature (under certain experimental conditions) about 17°C above the surrounding air; under the same conditions, but with the scales and hair rubbed off the thorax, the temperature fell to about 8°C above that of the air" (V. B. Wigglesworth, *The Life of Insects* [New York: World Publishing, 1964], p. 19).

Already toasty warm in their sumptuous fur coats, some moths can raise their body temperature further by vibrating their wings for several minutes before flying. The flight muscles grow several degrees warmer than the rest of the body, which itself is warmer than the ambient air.

Wigglesworth also refers to exceptional cases of insects that have successfully adapted to extreme temperatures. These include mosquito larvae in arctic regions, which remain active at 0°C (32°F), and larvae that manage to develop in the hot springs of Yellowstone National Park at 40 to 50°C (104 to

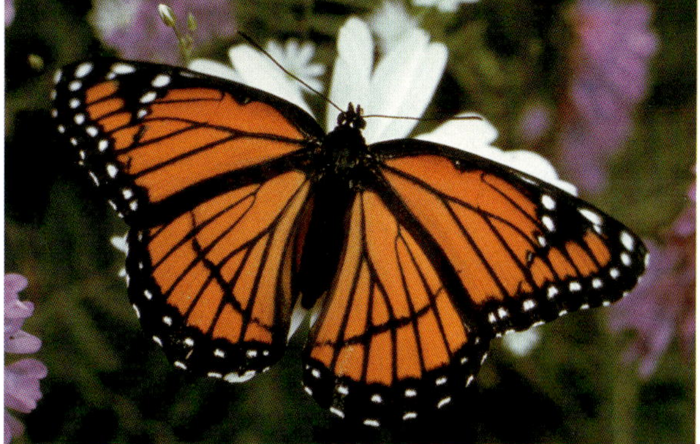

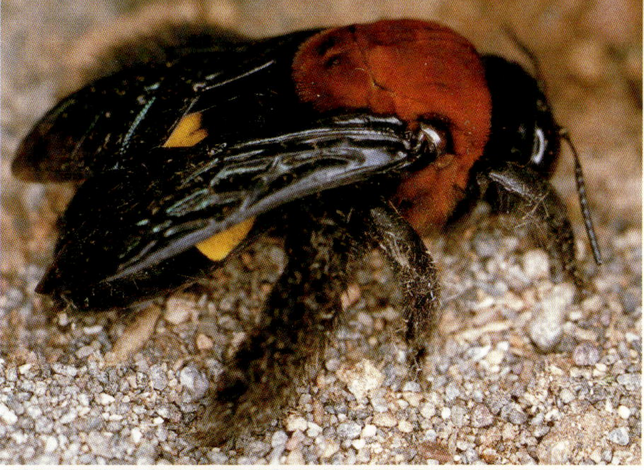

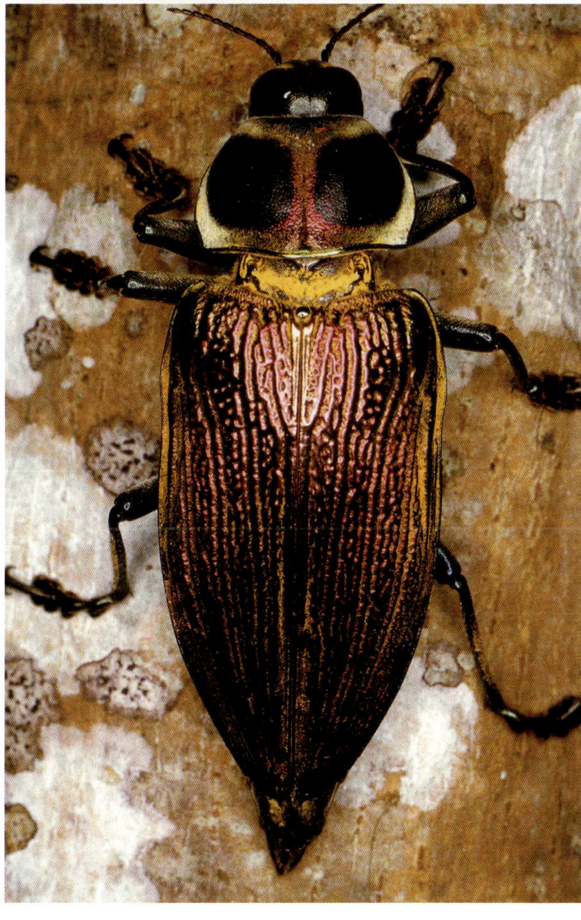

In the case of diurnal insects, heat and brilliant light provide optimal living conditions.
Top: The viceroy butterfly (*Limenitis archippus*, family Nymphalidae), a mimic of the monarch butterfly from eastern North America.
Middle: A large wood-burrowing carpenter bee (*Xylocopa* sp.) from the Amazonian region of Ecuador.
Bottom: One of the largest buprestids *(Euchroma gigantea),* a metallic wood-boring beetle from tropical America, measuring 5.5 cm. (2 in.).

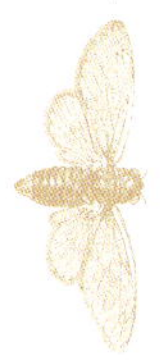

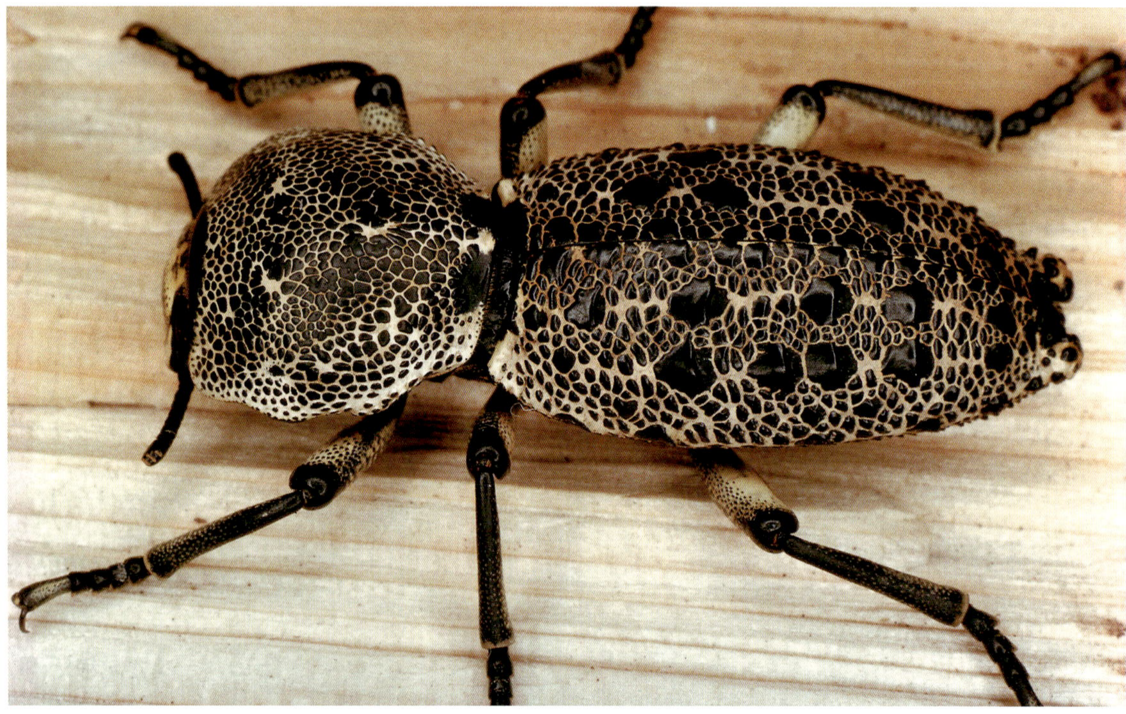

Seventy-five percent of the world's insect species are indigenous to the tropics, where they find ideal conditions that, except in desert regions, require few climate-related adaptations. They exhibit other types of adaptations, however, resulting from interspecific interactions. In climates where temperatures fluctuate widely, adaptations are primarily climate-related.

Above: *Zopherus marginipennis,* a member of the family Tenebrionidae from Central America, is admirably adapted to living in arid regions.

122°F). Between these two extremes, there are numerous less specialized insects that can also tolerate a wide temperature range, although they become lethargic at the upper and lower limits and die if exposed to more extreme temperatures.

The tolerance to cold exhibited by the underwing *(Catocala)* moths (family Noctuidae) found in temperate North America is also astounding. During one late September blizzard, I found two such moths that had been attracted by a light and landed right in the middle of the storm. Some tropical noctuids *(Ascalapha odorata* and *Thysania zenobia)* have been spotted in southern Canada, no doubt swept along by the warm winds that produce snow when they come in contact with a cold front. Although these powerful fliers are extraordinarily adaptable, accustomed as they are to both the cool cloud forests and the much warmer lowland rainforests, they risk exceeding the limits of their tolerance to cold when venturing so far north.

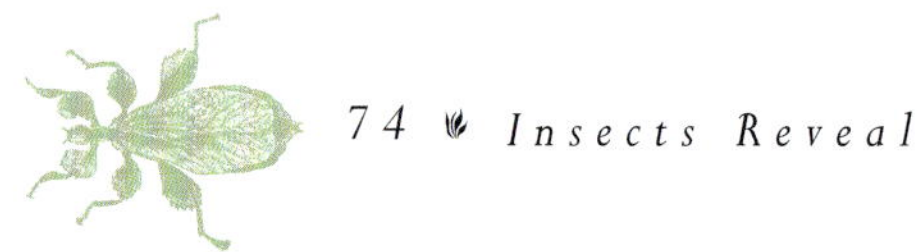

In contrast, large beetles in the subfamily Dynastinae (genera *Dynastes, Megasoma,* and others), although they inhabit the tropics, have little tolerance for daytime heat or direct sunlight, and are thus nocturnal. When attracted by the light of a lamp, these huge beetles sometimes fall onto their backs; if sunrise arrives before they have righted themselves, they will not live for more than an hour or two. At dawn, they normally return to their burrows in rotting tree trunks, where temperatures are lower than the ambient air. Long considered rare, the giant *Dynastes neptunus* comes to light traps only late at night, during the coolest hours.

Daytime or nighttime air temperatures may thus determine the periods when an insect is active, which must correspond with the species' physiological requirements. In the case of diurnal insects, heat and bright light together produce optimal conditions. Nearly all butterflies, tiger beetles, buprestids, and other beetles with metallic coloring are most active in warm conditions when their habitats are flooded by sunlight.

It is remarkable that many insects from temperate regions—where only the average temperature is "temperate"—develop as larvae in the heat of summer, enter the chrysalis phase in early fall, and spend the winter exposed to extreme cold. Others spend the winter as adults and re-emerge once the first warm days of spring arrive. The former are protected by cocoons, the latter by fallen leaves, thick snow cover, or other natural shelters. However, at temperatures of -25°C (-13°F), how much protection do these means actually afford?

In contrast to these adaptations to extreme cold, other forms of adaptation allow some insects to survive in very hot and arid regions. The defenses developed by these species are primarily anatomical and physiological, but they also behave in ways that allow them to exploit these characteristics to the fullest. These insects' most pressing need is to conserve water. They extract every last drop from their own excrement, and also prevent water loss due to evaporation by sealing off their integumental covering, thus becoming almost completely watertight. Tenebrionid beetles are the most remarkable and specialized of the these species. In the Namib Desert, along the southwest coast of Africa, some of these beetles climb to the tops of the sand dunes where they position their bodies almost vertically to catch and collect wind-borne moisture; they then drink every drop as it trickles down their bodies to their mouth parts. Others lift themselves up on their legs, which are exceptionally long by tenebrionid standards, thereby creating a space of a few millimeters between their body and the burning sand. Some tenebrionids are so watertight that I once had to place a specimen

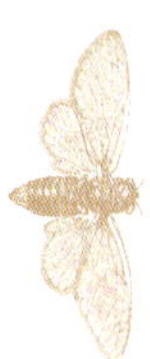

in a killing jar (containing potassium cyanide) for four successive nights because it proved so resistant to the poisonous vapors. Each morning, thinking that it must have died, I slipped it into a paper envelope, only to find at the end of the day that it was still alive. Any other beetle would have died in less than an hour. Once mounted and placed next to a stove to dry (in January, when I had just returned from an expedition to Costa Rica), the specimen took more than a month to harden. Under similar conditions, a different specimen of comparable size could have been removed from the spreading board after only a few days.

The adaptations just described have resulted from climatic changes that occurred over such long time periods that these species did not have to relocate; each new generation experienced only a gradual change. The process of adaptation followed a slow and gradual course, spread out over millions of generations. No matter how barren ecological regions may be, they always offer alternatives for adaptation. Slowly but surely, a natural process of give and take defines and circumscribes each species' ecological niche.

The operation of natural selection results in each species' developing its own "vocation," as adaptations in a particular direction tend to be favored in response to external forces.

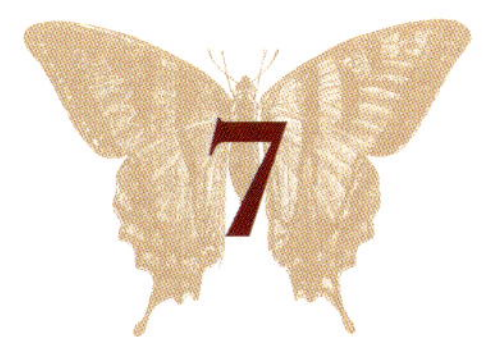

continental drift and
the spread of insects
throughout the world

In 1915, the German meteorologist Alfred Wegener published his theory of continental drift, which found few supporters at that time. Wegener was eventually proved right, and his views are generally accepted by earth scientists today.

Nevertheless, nonscientists have yet to come to terms with the idea that the continents have moved halfway around the globe over hundreds of millions of years. This is understandable because the changes described by Wegener have little or no bearing on what a person experiences in daily life. After all, what could be more stable than a continent?

Insects appeared on earth several million years before the continents began to drift apart; this fact helps explain various mysteries relating to the dispersal of the world's insect species. One could not possibly understand their distribution without recognizing that all of the continents were at one time combined into a single supercontinent, which Wegener named Pangaea (from the Greek word meaning "all-earth"), and that the continental masses began their drift across the oceans some 180 million years ago.

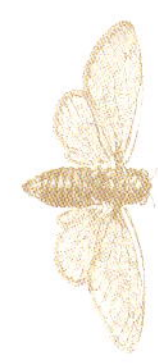

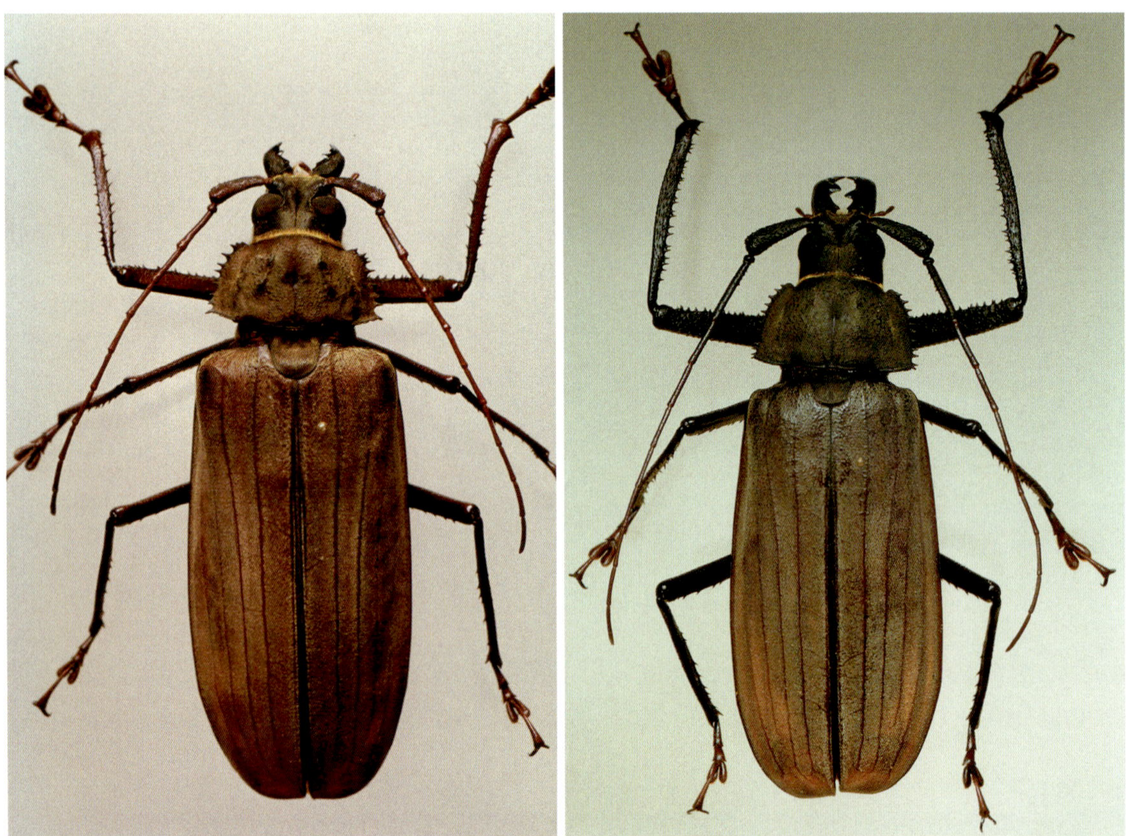

The ancient origins of these longhorn beetles (subfamily Prioninae, family Cerambycidae) predate the era of continental drift, when species were relatively undiversified. The Prioninae still exhibit little diversity. The two species shown have a common ancestor but are separated from it by millions of years and from each other by thousands of miles; their resemblance is so striking that one might think they were members of the same species. However, *Mecosarthron domingoensis* (left) is indigenous to Panama, whereas *Xixuthrus microceros* (right) is found throughout Southeast Asia.

In effect, Pangaea comprised those land masses that emerged over 225 million years ago. It included the continents we know today, although their contours were different and they were grouped in a way that we would hardly recognize. The land masses were clustered together, with South America and Africa fitting tightly to each other. The cluster then split into two groups. The one in the Northern Hemisphere—Laurasia—comprised North America, Greenland, and Eurasia, while the one in the Southern Hemisphere, a densely packed group called Gondwanaland, included South America, Africa, Antarctica, and Australia, all wedged together. Surrounding Africa were the land masses of Madagascar and triangle-shaped India, whose ensuing journey proved remarkable.

Pangaea's slow breakup did not occur concurrently throughout its various regions, and while some continents drifted either east or west, they also tended to shift northward.

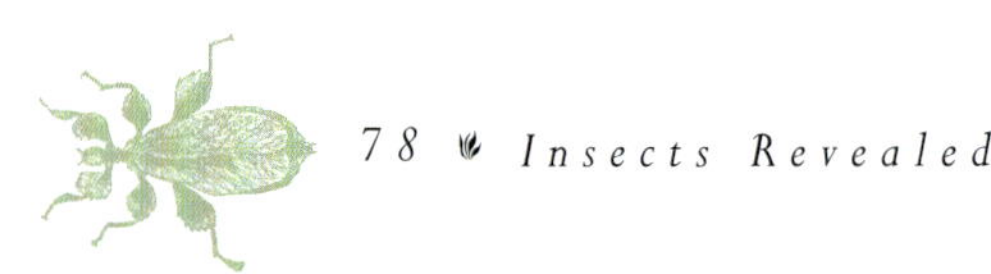

North America was the first continent to break away from Pangaea, drifting westward while also tracing a northerly arc. South America then began to move both westward and northward. In order to gain a sense of the distances covered by these continents in their travels, bear in mind that when North America was part of Pangaea, it was so far south that the equator passed through what is now Florida. The northern tip of South America lay almost on the equator; its northward drift was thus less pronounced than that of North America, corresponding to the present difference in latitude between Caracas, Venezuela, and Quito, Ecuador. Africa, which at one time showed a pronounced eastern tilt, became realigned along a north-south axis and, like the Americas, began moving northward. (At that time, the equator, which now runs through Gabon and Kenya, would have passed through Morocco.) This shift dragged along Madagascar, a piece of ancient Africa that gradually broke away from the continent. India followed a similar northward route, although it also began to shift toward the northeast, that is, toward Asia. It ended up colliding with the Asian plate with such force that the two continental plates penetrated each other; India thus became fused with Asia, and the Himalayas are living proof of the pressures unleashed by this cataclysmic impact.

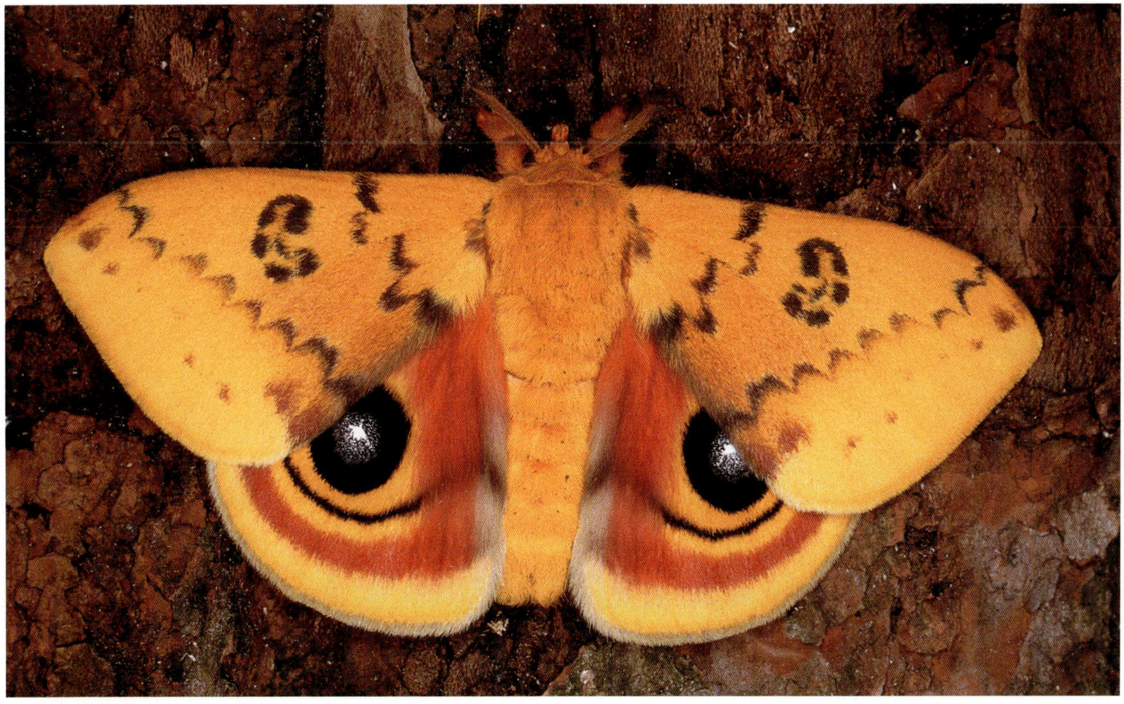

The io moth *(Automeris io)* is the only species of the genus *Automeris* found in our northern latitudes, and the only one to have adapted to North America's climatic changes. The genus is predominantly tropical; some one hundred other *Automeris* species are found in the warmer regions of Central and South America.

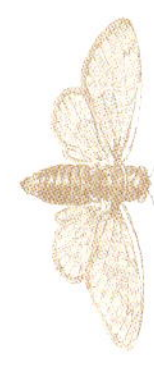

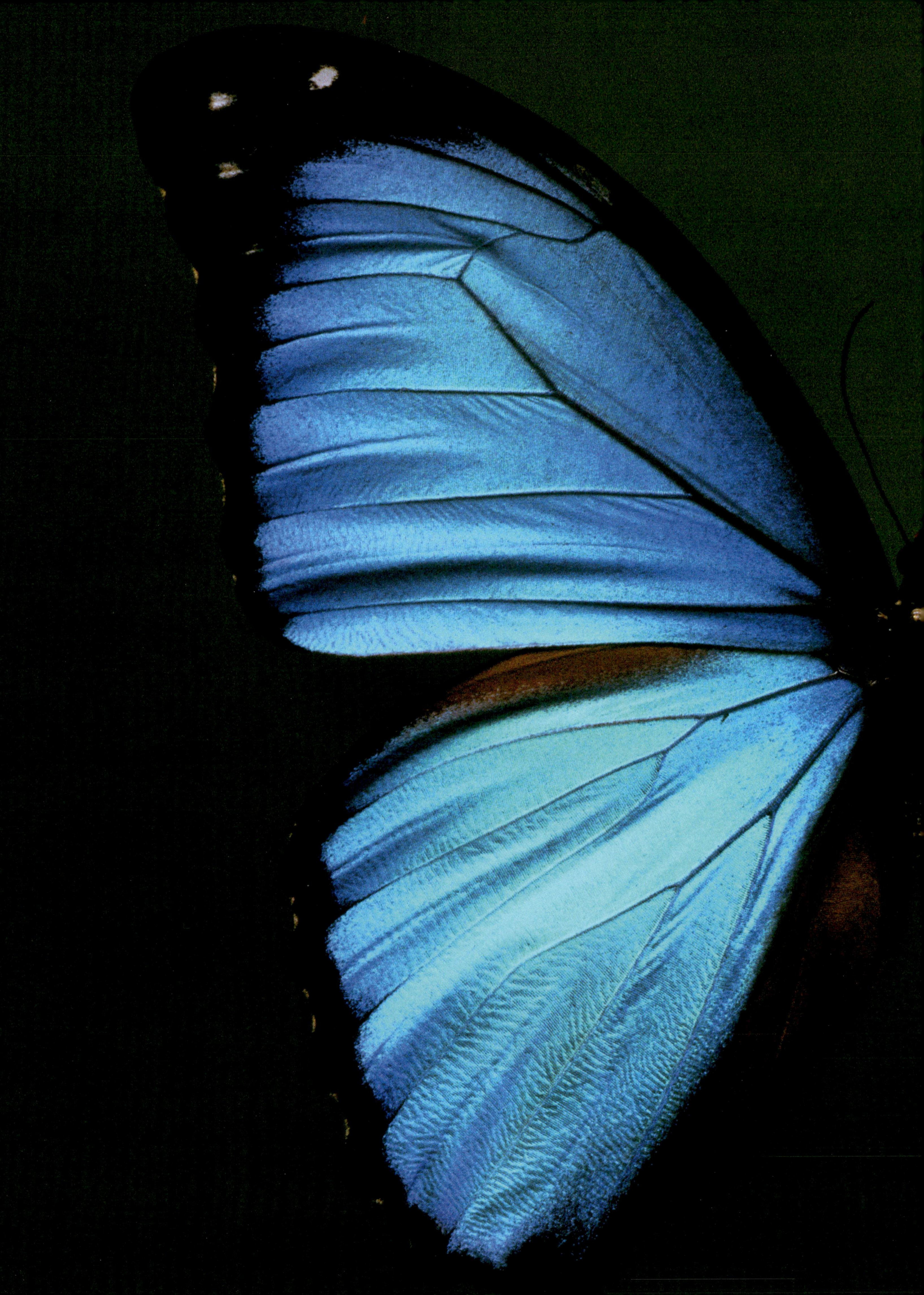

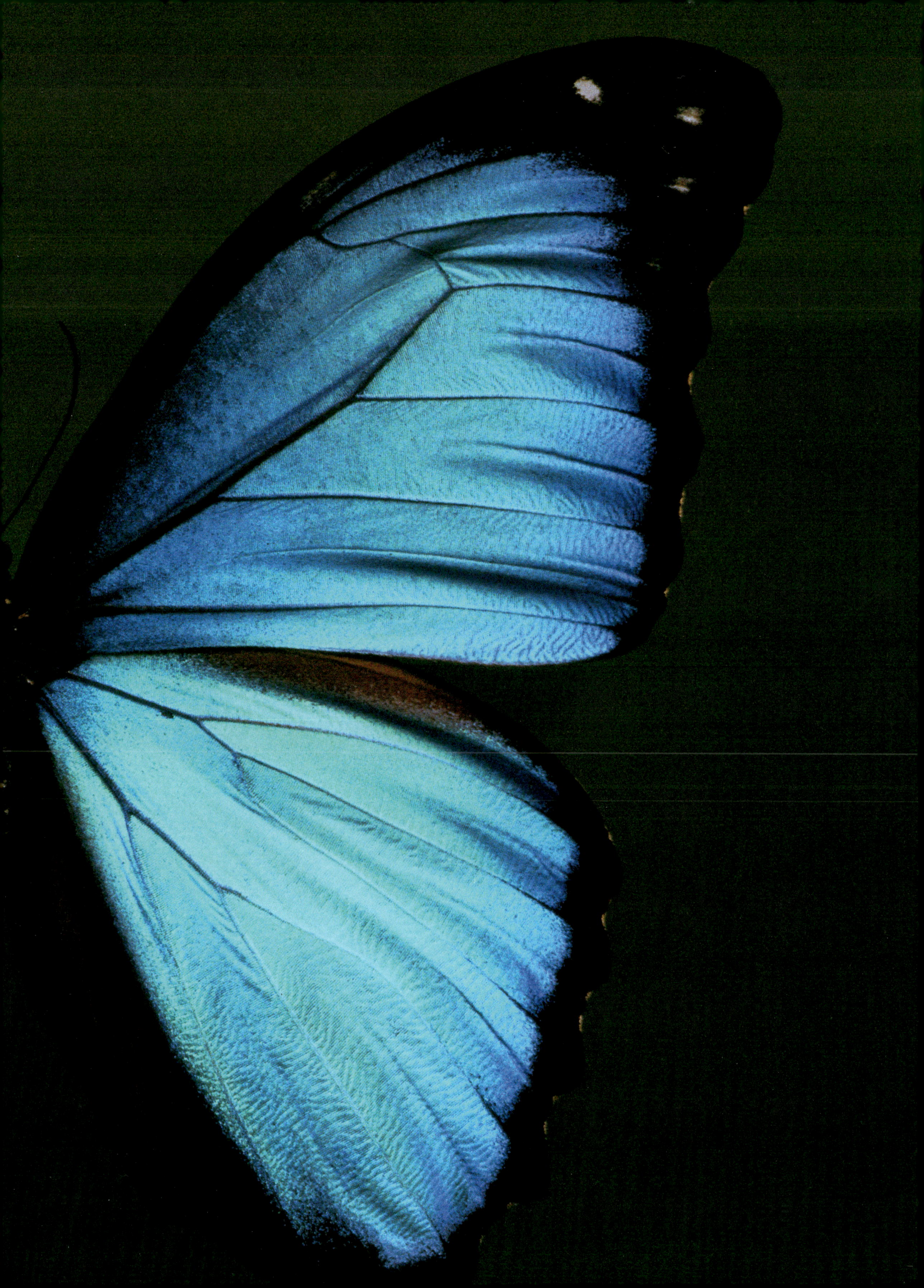

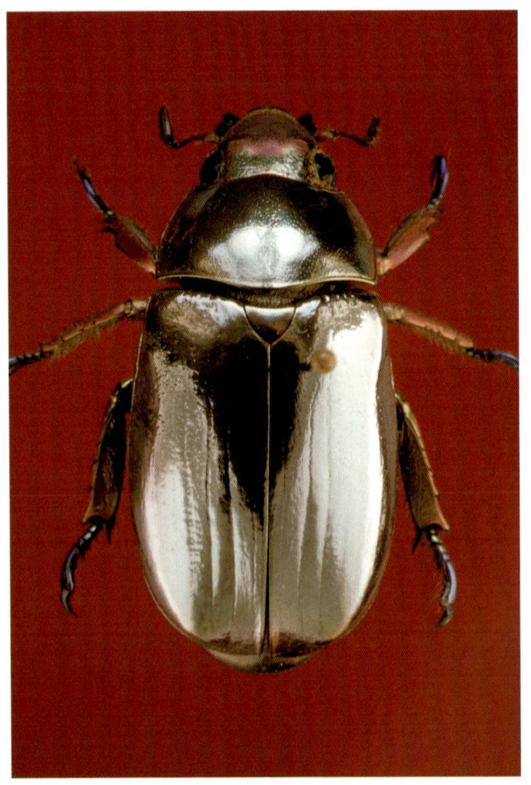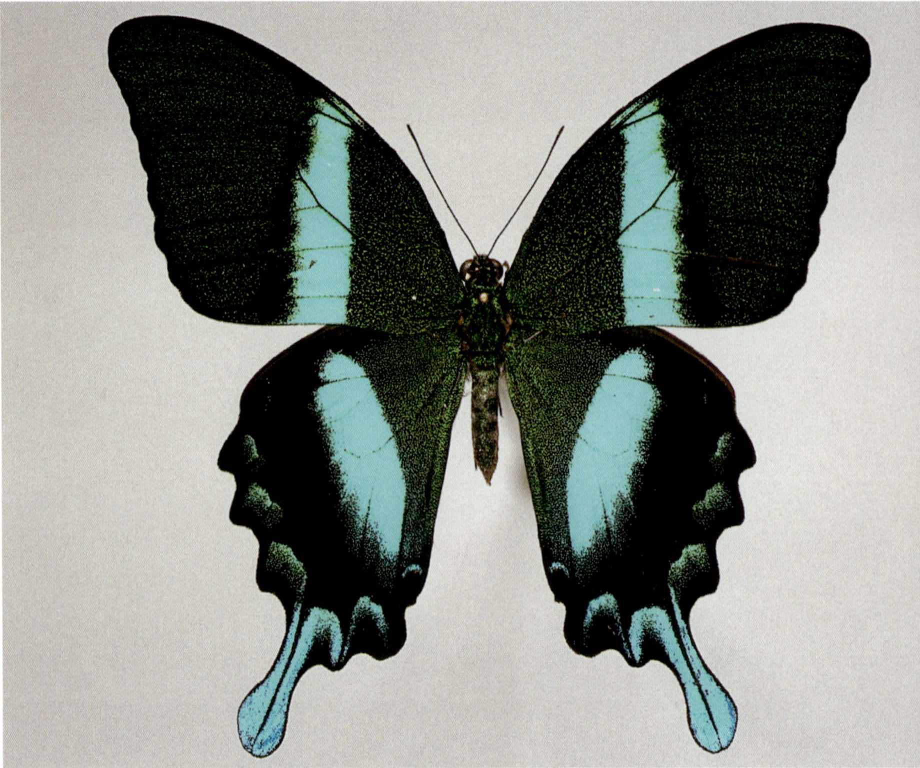

The distribution of insect groups and species around the globe is somewhat puzzling. Some occupy one or more continents; some live only on a single island; others occupy tiny areas no larger than a pinpoint on a map of the world.

Above: The numerous species of scarab beetles in the genus *Plusiotis* are found at moderate altitudes from Central America to Ecuador. Shown here at left, *Plusiotis chrysargyrea* from Costa Rica. The superb peacock butterfly (*Papilio blumei*, family Papilionidae) (right) is found only on the Indonesian island of Sulawesi. This island has always been isolated, which explains why it has many species that are found nowhere else.

During the redistribution of the continents of the Southern Hemisphere, Antarctica was the only land mass to travel south; Australia and New Guinea, which formed a single plate, moved toward the northeast.

What kind of powerful forces could have caused the continental plates to break apart and scatter around the world into the positions they occupy today? The answer lies at the bottom of the sea, within the deep trenches that crisscross the ocean floor. Lava flowing out of these trenches constantly forms new deposits, causing the seabed on either side to spread apart. The pressures exerted by the spreading seabed initially set the continental plates in motion, and this drifting process continues today at the

Preceding pages: The morpho butterfly (*Morpho didius*) makes its home in the Amazonian region of Peru. Several families of the order Lepidoptera, including the Morphidae, are indigenous to tropical America. However, most species from this family are found in well-defined sectors of this biogeographical zone.

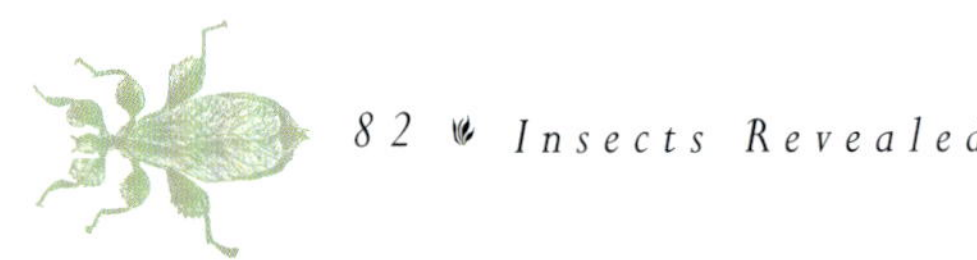

rate of several centimeters per year. These powerful forces are still at work and serve to keep the continents in a state of precarious equilibrium. In fact, the continents are subjected to opposing forces exerted by the ocean floor, which in some places has no choice but to slide underneath other land masses in a subduction zone, where it eventually melts within the earth's interior. None of this occurs gently, and these pressure points bring about the volcanic activity that threatens many parts of the world, most notably Southeast Asia, the Andes mountains, and the entire west coast of South America.

The unfolding of these dramatic events provides the vast setting for tracing the movements of tiny animals. Insects may have conquered the world, but their success can be ascribed to the advantages they enjoyed by arriving on the scene when Pangaea's land masses were still interconnected. The supercontinent provided insects with opportunities that they exploited for one hundred million years before the continents began to break apart. The climatic conditions at the time, which were generally warmer and wetter than they are today, also served to further insects' development and expansion. When Pangaea finally broke apart some 180 million years ago, insects belonging to all of the orders that we are familiar

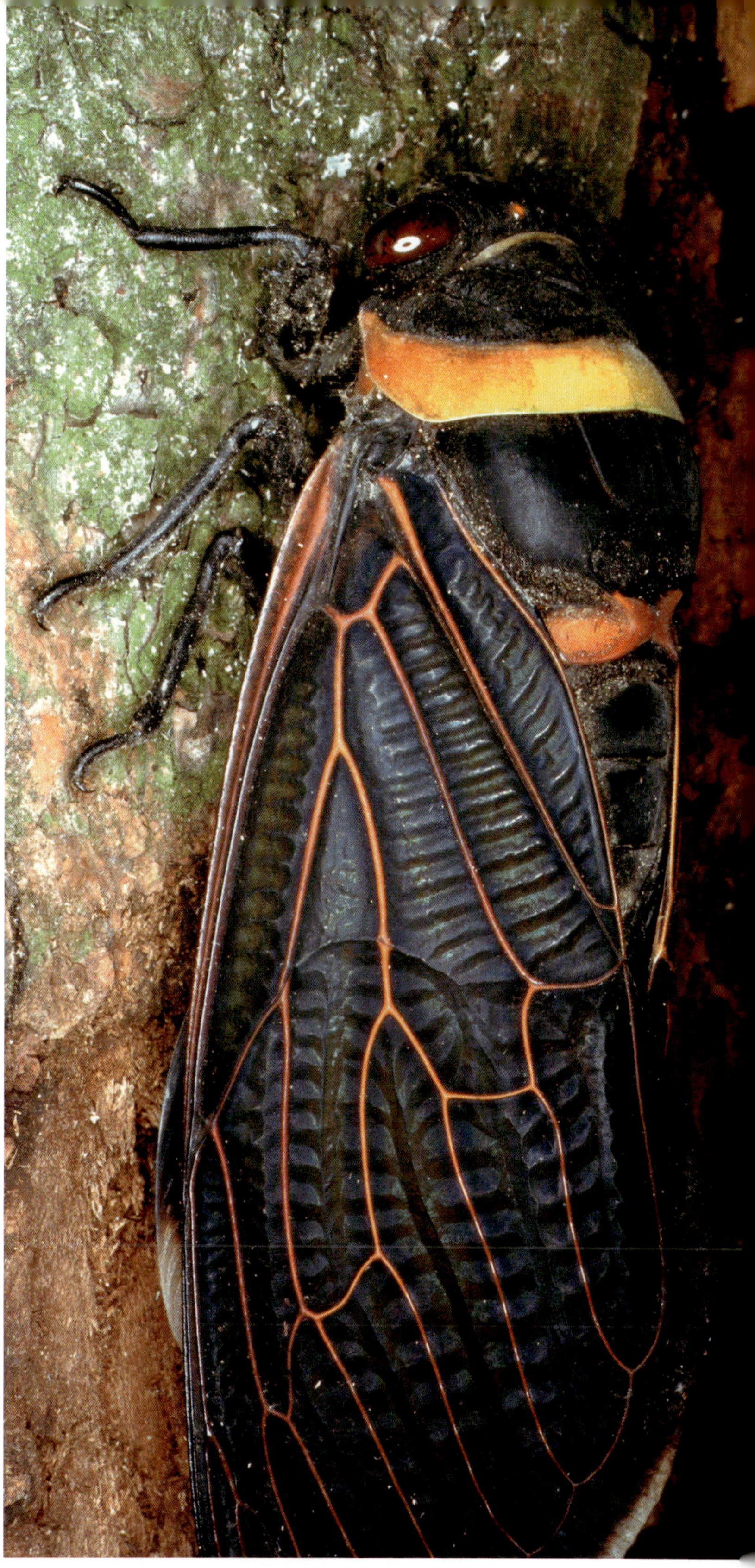

Tacua speciosa, a cicada with wings as colorful as those of butterflies, inhabits Southeast Asia, one of the world's richest zones for insect species.

with had already spread across the continents. This largely explains the extraordinary similarities of closely related species that now inhabit different regions, separated by huge distances across the oceans. These species almost certainly had a common ancestor that once inhabited Pangaea.

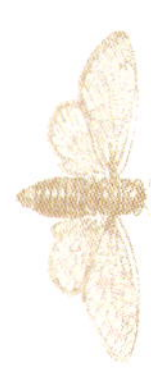

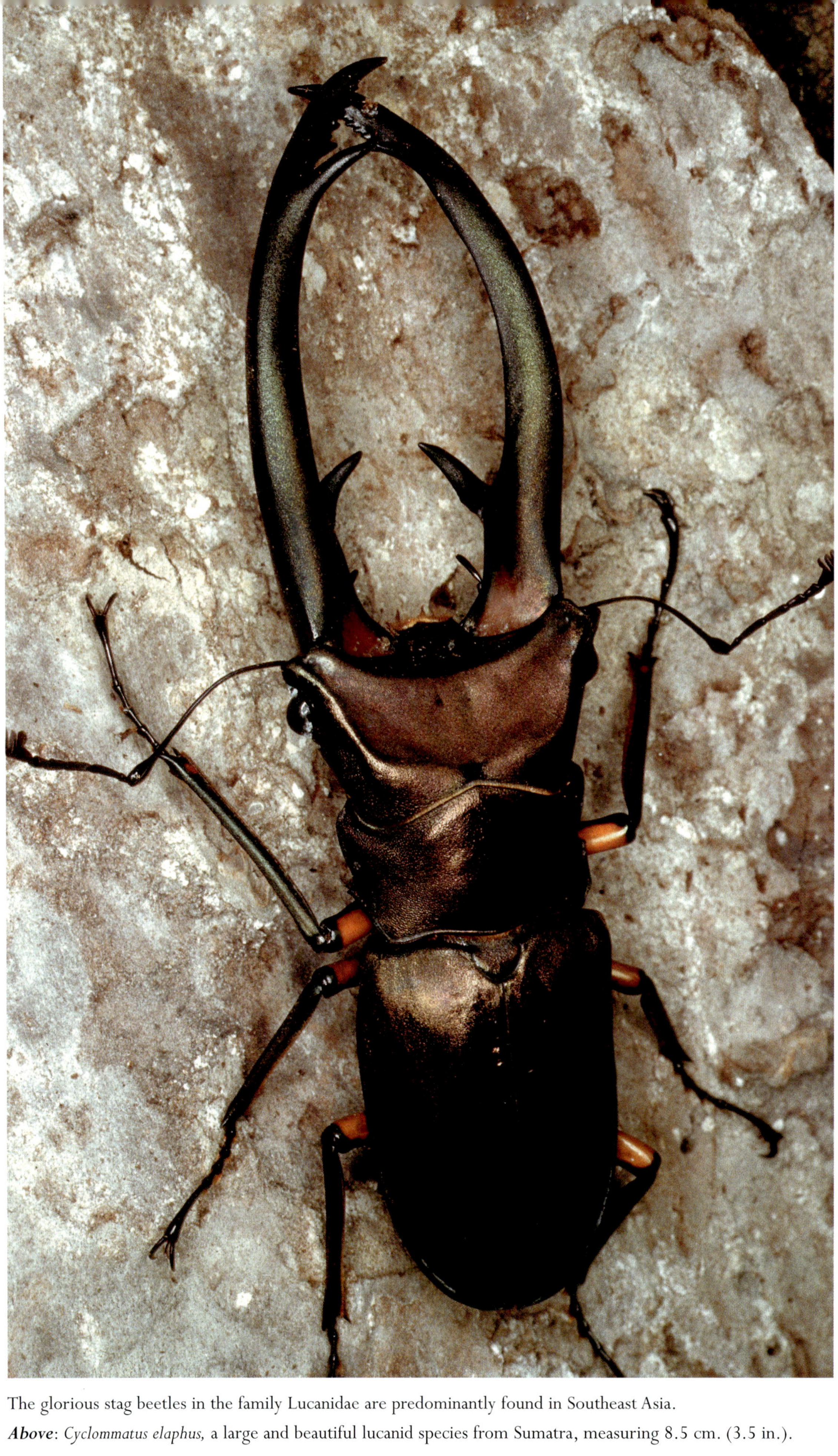

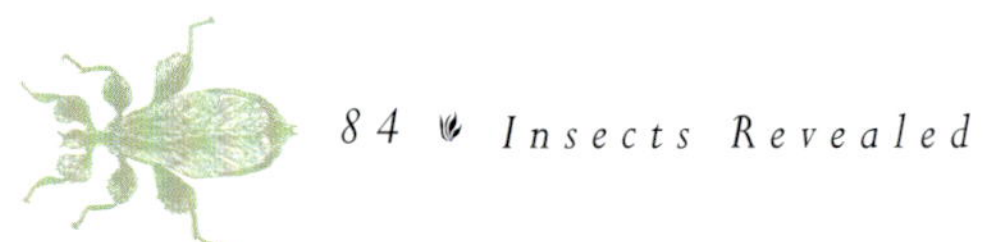

The glorious stag beetles in the family Lucanidae are predominantly found in Southeast Asia.

Above: *Cyclommatus elaphus,* a large and beautiful lucanid species from Sumatra, measuring 8.5 cm. (3.5 in.).

When the continents broke away from the supercontinent, the natural land bridges that had enabled species—even those with limited mobility—to circulate freely were broken. Adrift on the oceans, insects on the newly isolated continents evolved independently, since contacts with populations on neighboring continents were no longer possible. The continents' movements over millions of years brought about considerable climatic changes in some regions; consequently, global temperature differences became more pronounced than they were during the Carboniferous era. Insect populations on each continent thus either had to adapt to local conditions or modify their geographical distribution within the land mass. This served to differentiate the insect fauna of the world's various biogeographical zones, which, in the beginning, were much more homogeneous.

Although continental separation ripped apart numerous land bridges, new bridges were formed several million years later when previously isolated continents drew closer together during the slow drifting process. The Bering Strait, which separates North America from northeastern Asia, became a natural bridge for the largest and most recent migrations of insects, plants, animals, and humans. These migrations were aided by the drop in sea level during the Pleistocene epoch, some three million years ago. Vast quantities of water were absorbed by the icecaps that formed over large portions of North America and Europe; as a result, parts of the seabed that until then had been submerged under shallow water were left exposed.

The drop in sea level also uncovered many shoals and shallow places that have subsequently been reclaimed by various bodies of water. The Torres Strait, which separates Australia and New Guinea, was at one time dry land; hence the striking similarities of the flora and fauna of the two regions, which are now separated by 200 km. (125 miles).

At one time, the Sunda Islands, which extend from Sumatra to Bali, were not islands as we know them today in our interglacial period. These areas of high relief bordering Indochina became part of the continent when sea levels dropped more than 100 meters (330 feet) during successive glacial periods. The exposed seabed between the islands and the continent thus became the lowlands of a vast peninsula that included the Malay Peninsula, Sumatra, Java, Borneo, and the Philippines; this represented an enormous increase in Southeast Asia's surface area.

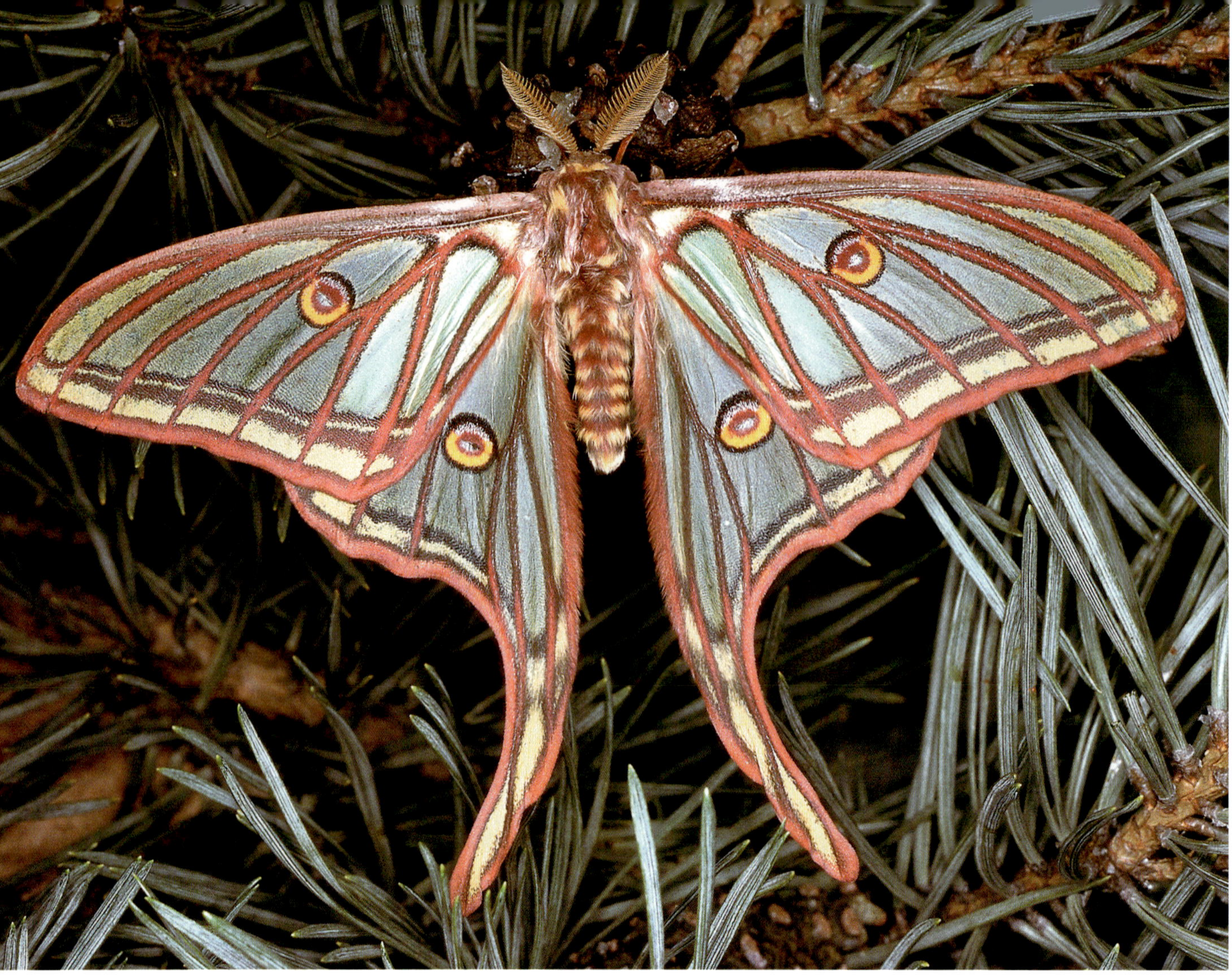

The Earth's surface has been in constant flux throughout the various geological eras. Some insect populations have reaped benefits by extending their range, whereas other species have seen themselves reduced to a single hilltop or far-flung valley. Such isolated species are known as stenotopic (from the Greek *stenos,* narrow, and *topos,* place). Here is one remarkable example.

Above: The Spanish moon moth *(Graellsia isabellae),* a magnificent European saturniid that is the only species in the genus. Long ago, it may well have had a more extensive range. Today, it is limited to several isolated sites in the pine forests of southwestern France and northern Spain. This moth is related to those of the *Actias* genus, including the luna moth *(Actias luna)* of North America and some Southeast Asian species. *Graellsia isabellae* has five subspecies, including *paradisea* (shown), from Herona province in Andalusia, Spain.

It comes as no surprise to discover that so many closely related insect species are found throughout the island-studded Southeast Asian region, as if it formed a single land mass. These modern species have common ancestors, which once reigned over the Indochinese peninsula when its entire surface area was exposed to the air. Nowadays, species on the various islands exhibit characteristic variations in form that point to a past when continental and island environments alternated. As Charles Darwin discovered in the Galapagos Islands, nothing promotes the evolution of new forms and species more than geographical isolation, especially in island settings.

Many other geographical changes occurred in coastal regions when the sea level dropped during the glacial periods and rose during the melting of the ice caps. For

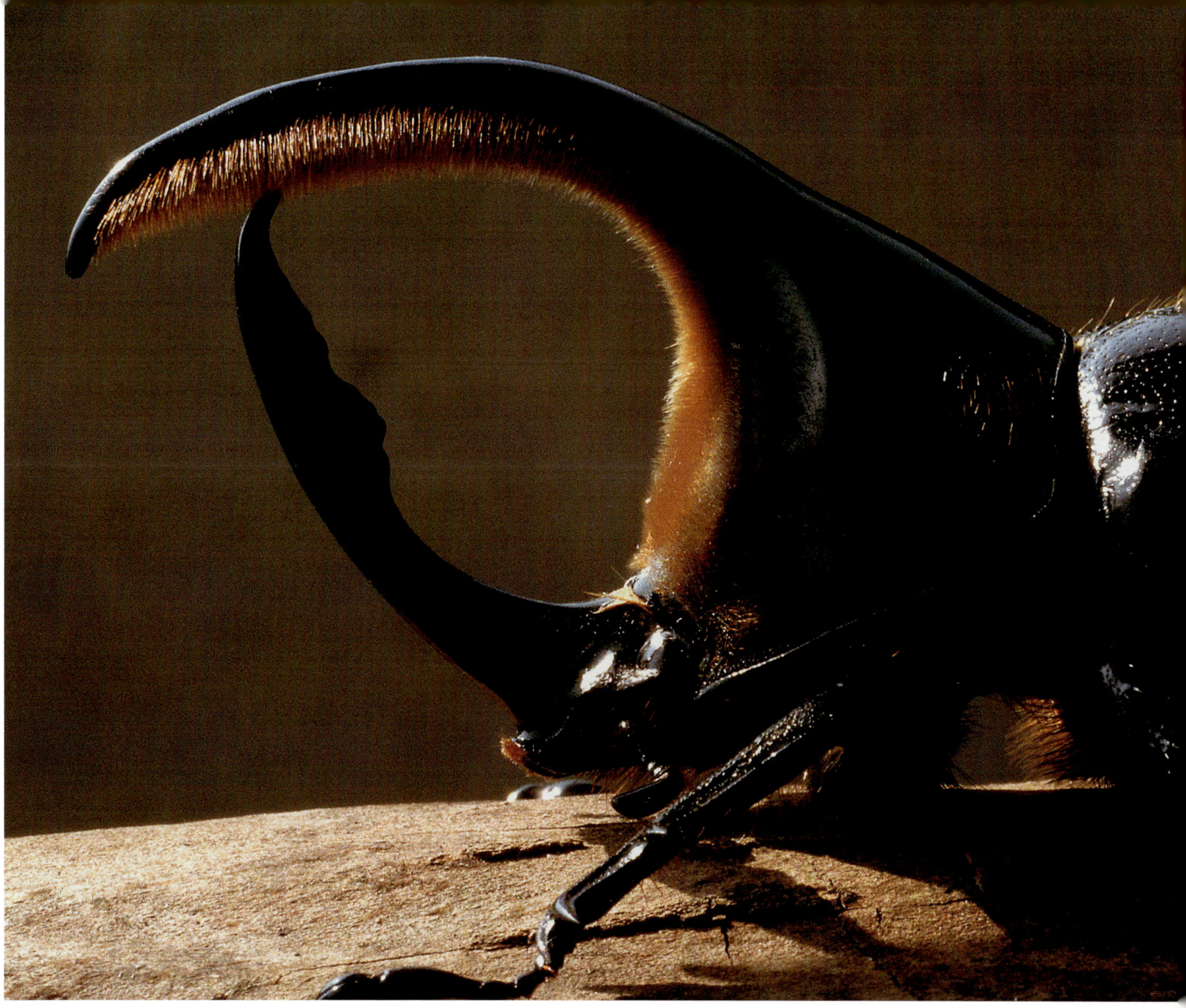

The first specimen of this large scarab beetle, *Dynastes satanas* (subfamily Dynastinae), was captured and described in 1909. In the 1980s, a few specimens were found in Bolivia. This insect is renowned for its rarity. Little is known about such rare species, and this lack of knowledge contributes to their rarity.

The only pictures I had ever seen of this beetle were published by Gilbert Lachaume in the series *Coléoptères du monde* (Beetles of the world). I was deeply moved to photograph this superb male specimen, lent to me by a friend as if it were an everyday kind of event.

example, Trinidad was not created by the volcanic activity that produced the other islands of the Lesser Antilles; it is a former peninsula of Venezuela that was cut off from the continental mainland by rising water levels. Trinidad's rich insect populations thus attest to its former link to the South American continent.

The climates of the world's continents have become drier or wetter owing to these geographical changes, which affected both inland and coastal areas. In the Amazonian basin, the boundaries of the vast rainforest have receded at times, forming small oases of lush greenery surrounded by areas of forest adapted to drier climates; when humidity levels rose, the rainforest expanded once again. As a result, animal life was forced to adapt to changing climates, in particular the insect fauna, whose unparalleled diversity today reflects the variable conditions of the past.

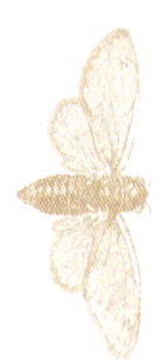

Until the beginnings of the Pliocene era some 12 million years ago, the two Americas had continued to drift without coming into contact with each other. At that time the Isthmus of Panama began to emerge, first as a series of isolated volcanoes in the sea, then as a strip of unbroken land linking the two continents. Numerous plant and animal species crossed this land bridge, traveling both north and south. However, since tropical America had many more insect varieties, North America gained more species than it lost, with the exception of mammals.

Because these population exchanges occurred over very long periods, it is not always possible to determine a species' origin. However, flying insects belonging to the family Sphingidae (sphinx and hawk moths), several species of which are indigenous to both North America and large regions of tropical America, probably traveled from the south to the north rather than vice versa. *Automeris io* (family Saturniidae) is the only species of that genus found in the northern latitudes, and may provide a glimpse of a time when North America was warmer; the one hundred or so varieties of *Automeris* indigenous to tropical America show a distinct preference for warm climates.

In considering the many changes that insects have undergone during geological time, one finds many aspects of their current distribution patterns that raise difficult questions. Why, for example, are large families of butterflies such as the brassolids (owl-imitating butterflies) of tropical America and the Amathusidae of Southeast Asia restricted to their respective regions even though they have so many traits in common? Did they evolve in isolation after having originated from a common ancestor long ago? Or might this be a case of convergent evolution?

Other families, however, such as the Morphidae (morpho butterflies), Heliconiidae (heliconian butterflies), and Ithomiidae (clear-wing butterflies), which are exclusive to tropical America, have no counterparts anywhere in the world. Certain families or subfamilies that dominate given regions are represented elsewhere, but only in a minor role. Stag beetles of the family Lucanidae are some of the most glorious examples of Southeast Asian insect life, but they have only a discreet presence in equatorial Africa. Surprisingly, only certain rare species of Lucanidae are indigenous to tropical America, and almost all of these are primitive varieties, concentrated primarily in the temperate latitudes of Chile and Argentina.

The last places to be colonized by insects were the oceanic islands that were formed fairly recently in geological terms. Colonization depends on the distance between islands and the nearest continent; it also depends on winds and currents that carry vegetation to the islands, creating semi-hospitable environments for any animal species that might accidentally wash up on shore. Over thousands or millions of years, many newcomers that were propelled across the oceans by storms, favorable winds, or rafts of living plants attempted to germinate, take root, or otherwise gain a foothold on soils that more or less suited their needs. It is this "more or less" that filtered out those species with some chance adaptations that enabled them to survive in their new habitat—a tiny number.

The physical dimensions of islands play an important role as well. Larger islands have more varied contours and provide colonizing species with more diverse biotopes. As a general rule, however, the insects that manage to establish themselves on oceanic islands represent only a fraction of those species found on the continent of origin. In addition, transplanted island species are often smaller. This dwarfism is easily explicable when one visits islands such as the Galapagos and witnesses the tough and arid environments that insect immigrants faced.

Nevertheless, islands do not always have such punitive effects on the species they attract. After a while, the blanket of vegetation on some islands, such as Dominica, in the Lesser Antilles, may grow to luxurious proportions. Even if insect life on Dominica and its neighboring islands is not as varied as on the continent, it does include the most herculean examples of *Dynastes hercules,* a huge horned scarab beetle. Those from Dominica are among the largest and most beautiful found anywhere.

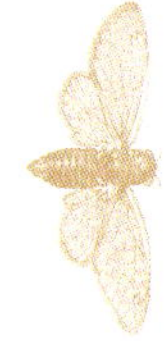

Two beetles and their wide-reaching influence on a passion and on this book

As a child, I once received a gift of two beetles from tropical America. These rather decrepit specimens (a *Golopha* and a *Psalidognathus*) held a special place in my heart for some forty years, as symbols of a promised land where there might be ten times—maybe even one hundred times—as many natural wonders as in North America.

I finally set off one day for the southern regions in the hopes of finding live beetles that might be distantly related to my two specimens. My personal collection includes a fine series of *Psalidognathus,* but I never did find a live one. As for the *Golopha,* I was lucky to spend an unforgettable afternoon in their midst in a Venezuelan cloud forest.

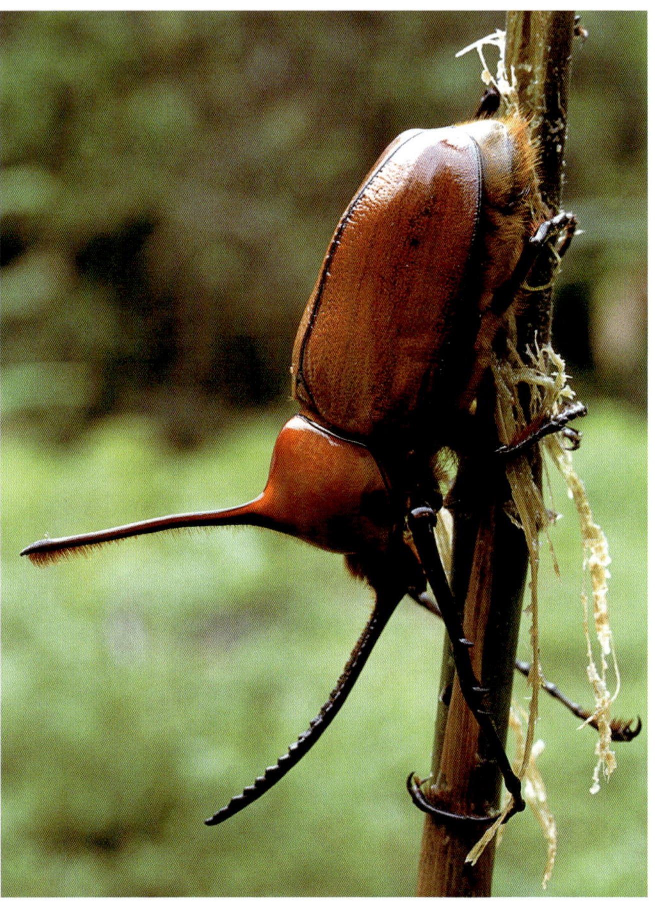

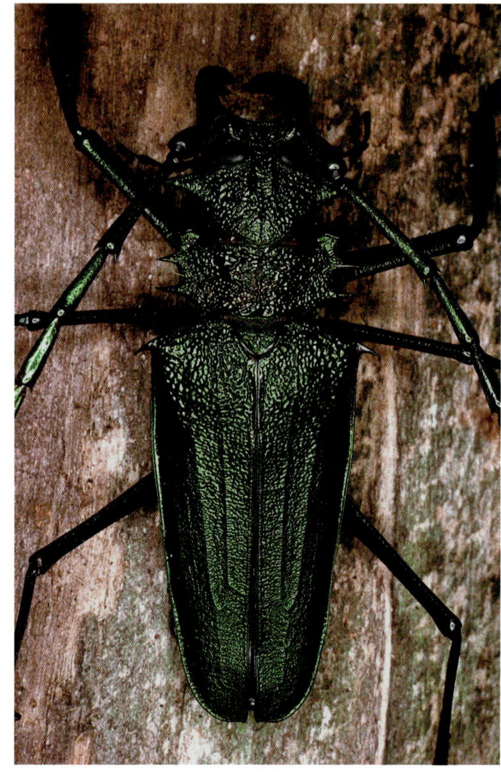

Left: A scarab beetle (*Golopha porteri,* subfamily Dynastinae) widespread in Colombia and Venezuela and restricted to areas where a suitable species of bamboo is found. This beetle is diurnal and is primarily active in misty, dimly lit regions.

Right: This magnificent longhorn wood-boring beetle (*Psalidognathus friendi,* subfamily Prioninae, family Cerambycidae) from Colombia is distinct from most of its relatives, which are generally dark colored and relatively undifferentiated.

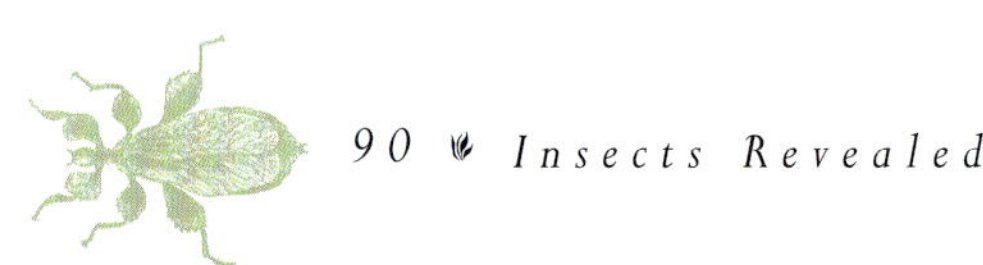

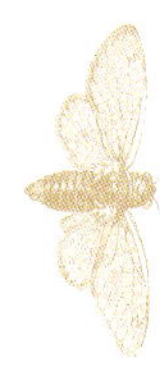

the evolution of insects
in tropical regions

The world's rainforests did not suffer directly the destructive effects of the glacial periods. Thus, their insect populations developed over millions of years with fewer interruptions than occurred in temperate regions.

Evolutionary processes have thus had free rein in these regions, and even insignificant traits have developed to extravagant levels.

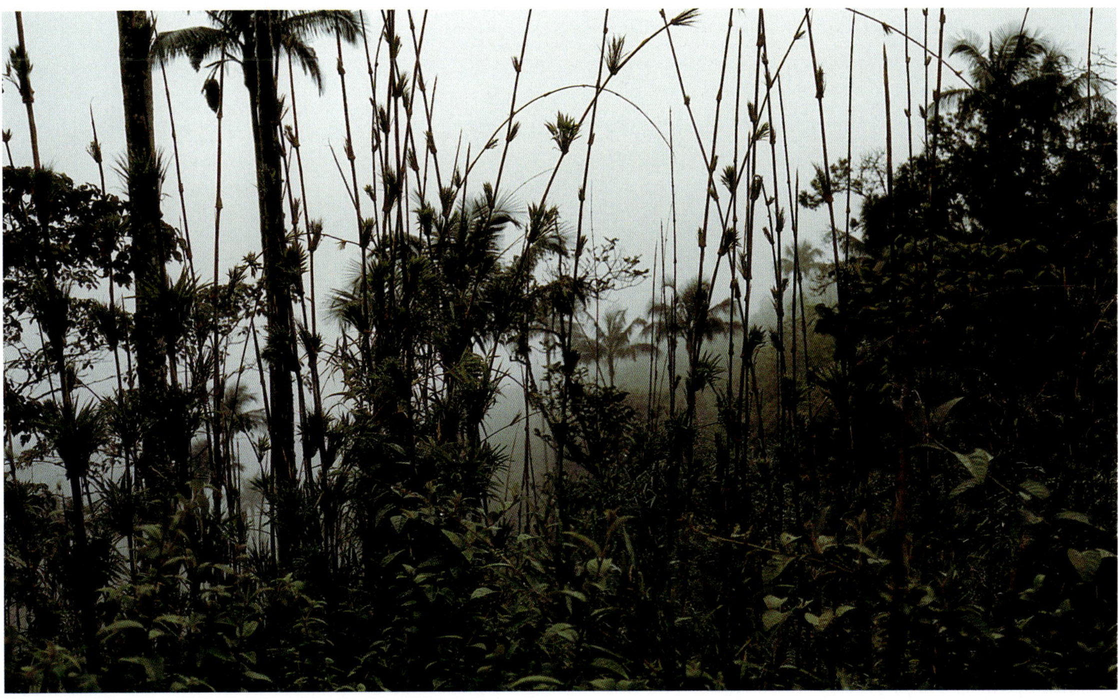

The typical habitat of the scarab beetle *Golopha porteri* in Venezuela.

In contrast, insects in temperate zones have evolved in a climate that imposes a rigorously defined periodicity. Given that each developmental stage is associated with a particular season, the biological cycle of any individual that fails to hatch at the most propitious moment will be seriously impaired.

The lush tropical regions have a more uniform climate—an endless summer that has lasted for millions of years—and are thus able to offer their insect fauna optimal conditions. The gigantic proportions and the fantastic colors and shapes of many tropical insects are testament to the favorability of these conditions.

In my youth in Quebec, we had a simplistic view of the tropical world since virtually no one we knew had ever been there. Any mention of the "warm lands" would lead us into an imaginary universe. References to hot climates always seemed to include the "jungle," or, more accurately, the humid tropical forest or the rainforest.

For many people, this biotope conjures up images of an extremely hostile environment, home to creatures that one would best avoid. However, one cannot speak of a tropical forest without specifying a particular type: a low-altitude humid forest, a seasonal deciduous forest, or even a cloud forest, which is itself divided into several distinct types based on elevation.

Although they are all tropical, these forests show enormous differences in their structural composition and their varieties of plant life. The associated fauna are also specific to each forest type. In the tropics, forests of various types may border each other or even overlap, depending on location, elevation, soil, and other factors such as winds and marine currents.

Consider the great humid forests of the low-lying regions, where temperature and humidity vary little throughout the year. The term "endless summer" is particularly appropriate for this biotope, which differs so radically in this way from temperate-region biotopes.

Opposite page: The author at "home" in his favorite place, the rainforest. Rainforests are sustained by high levels of heat and humidity. The species that live there have adapted to accommodate the enormous amounts of rain that fall daily. This milieu is certainly forbidding, but, as in other places, danger has its limits; if you are aware of them, you can live with it.

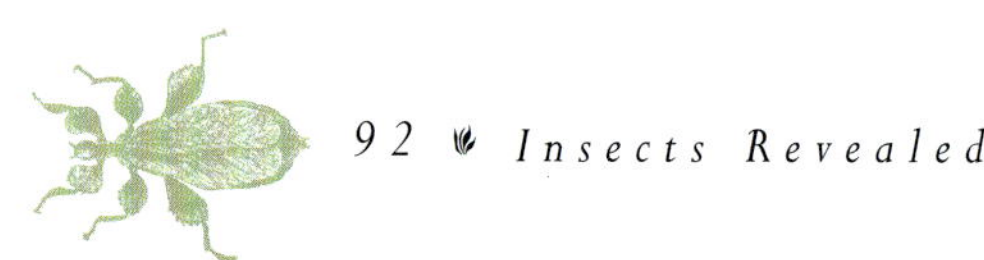

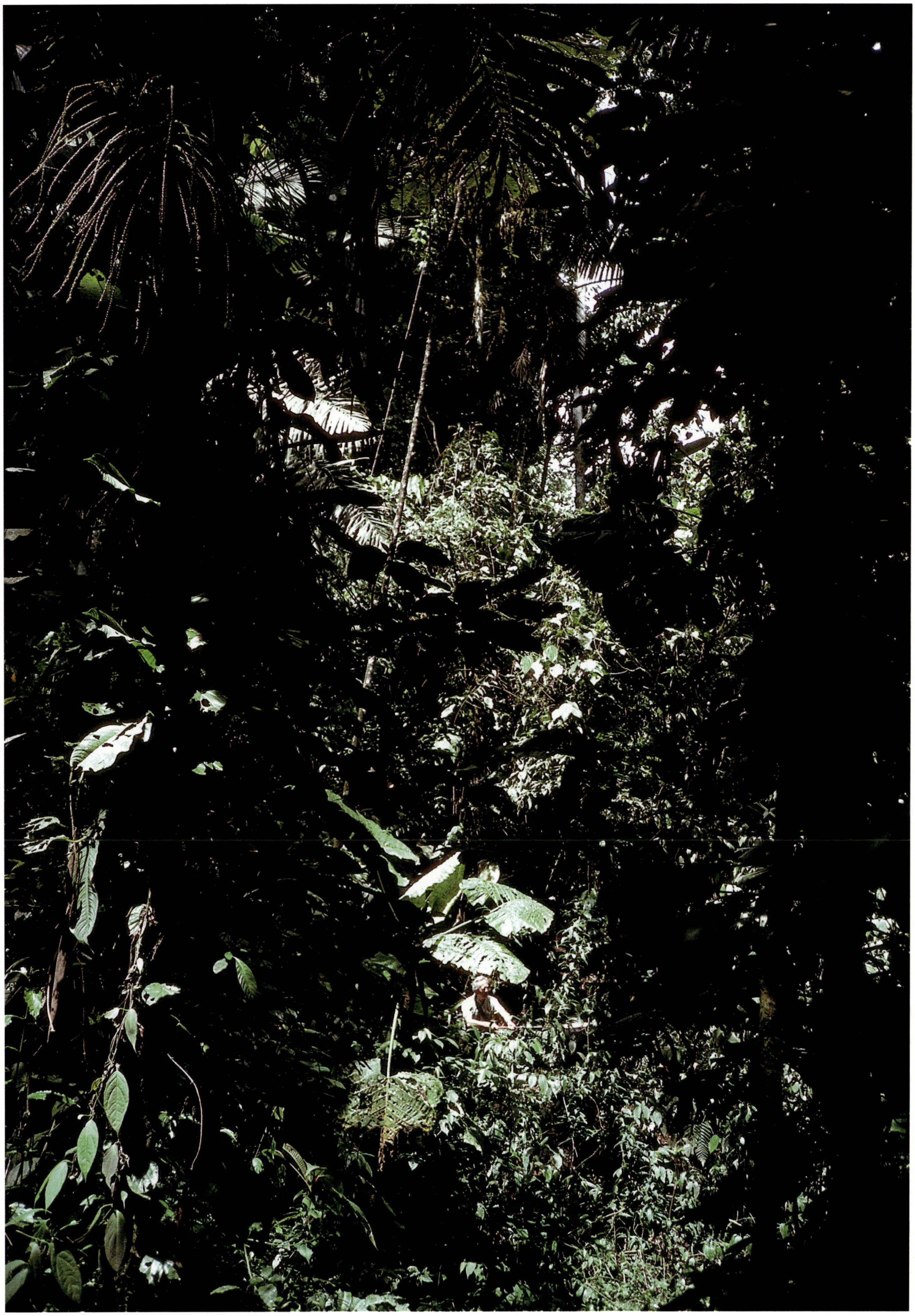

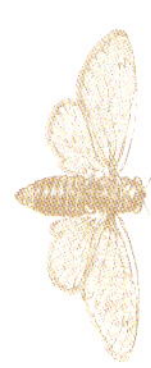

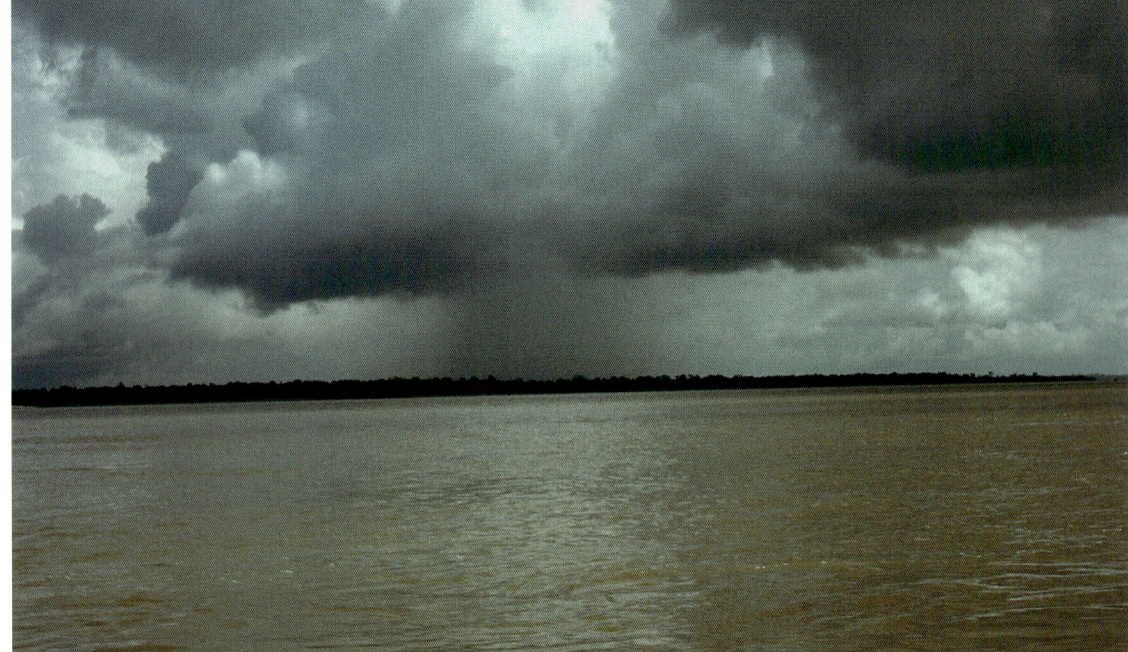

Rain on the Amazon.

How have insects evolved in this endless summer that, in many respects, is like the climate that prevailed in the Carboniferous era, when the insects first appeared? Such conditions have enabled insects to proliferate more rapidly than they have anywhere else. However, despite the extraordinary biodiversity of the tropical world, insect collectors visiting from the northern latitudes may be disappointed at seeing only a small proportion of the species described in field guides.

Faced with this puzzle, we may come to appreciate the practical knowledge that local entomologists have of their environment. Endless summer is an illusory concept for these specialists, who realize that there are almost as many seasons as there are hidden species of insects and of the plants on which they depend. To be sure, large numbers of species reproduce on an annual basis, but most reproduce only intermittently, in response to subtle variations in moisture levels in the immediate environment. Thus, although rich in species—75 percent of the world's insect species are found in tropical regions—rainforests may seem relatively devoid of insect life, depending on the area or time of day. This may seem improbable until we recognize and understand certain covert games that play out beneath the thick forest cover.

The deceptive rarity of observable rainforest insects is not due primarily to seasonal and transitory factors. In effect, it is a direct result of the huge number of species living within this ecosystem.

The overwhelming numbers of plant species in rainforests are structured in multiple levels (or strata) that experience decreasing amounts of sunlight as one

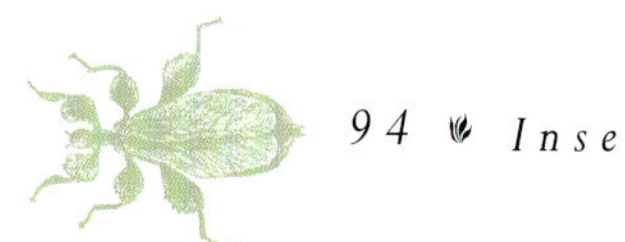

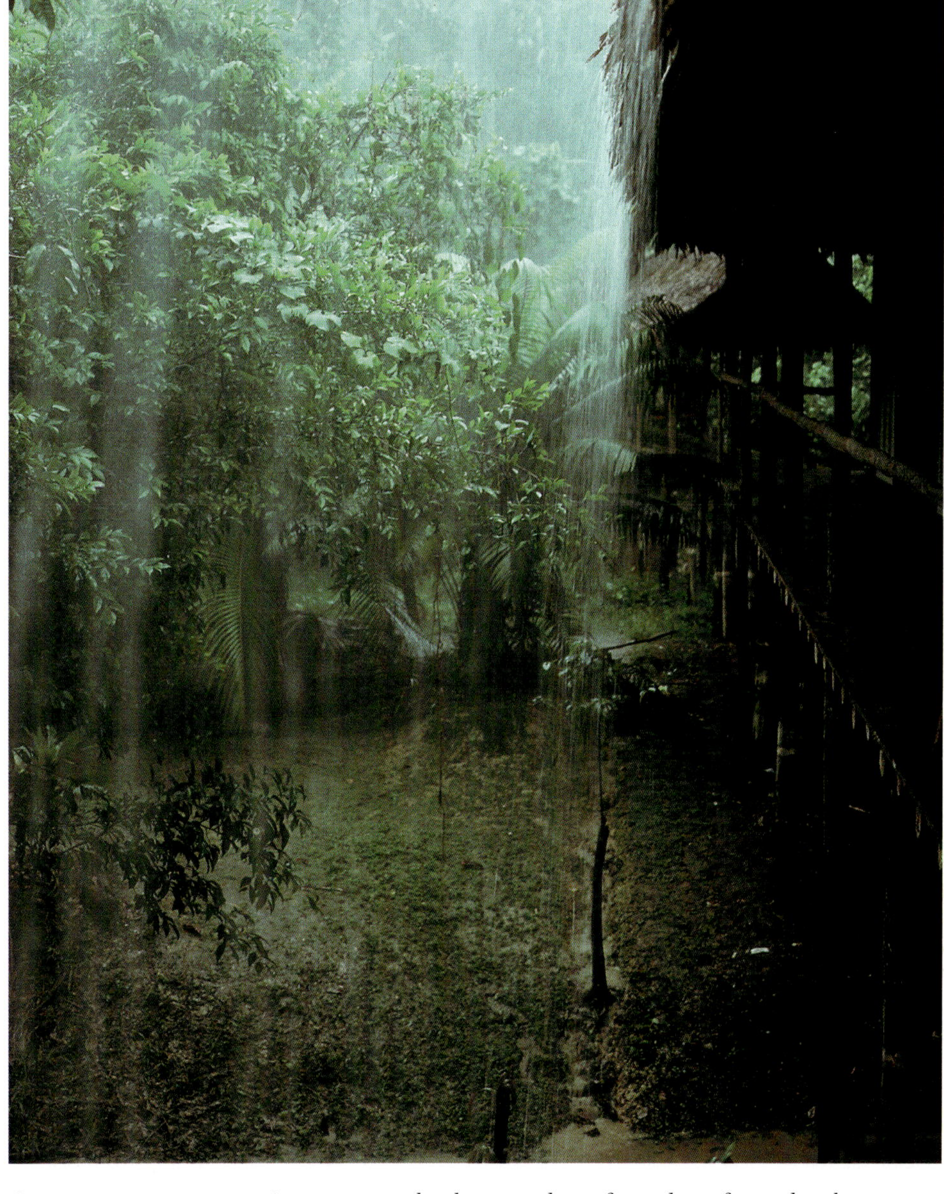

One ten-minute storm in Amazonia may be the equivalent of two days of rain elsewhere.

descends toward the forest floor. Some animal species live entirely within a single zone; others, depending on the time of day or night, move freely up and down the various strata.

Insect lovers from temperate climates who stroll along the paths of the forest floor—the darkest and most humid zone—may well meet with disappointment. They will see only diurnal species that have adapted to living in relative darkness, together with a few well-camouflaged nocturnal varieties.

So where do we find the legendary forest insects that, by all accounts, are so abundant in this habitat? In fact, they are dispersed from the forest bottom all the way up to the brighter and better-ventilated upper levels, where the many flowering trees hum to the sound of insects gathering nectar. One does not find these species in the lower levels, where there are few flowers.

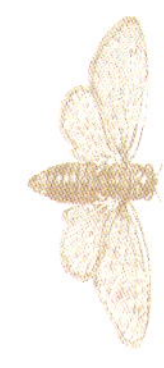

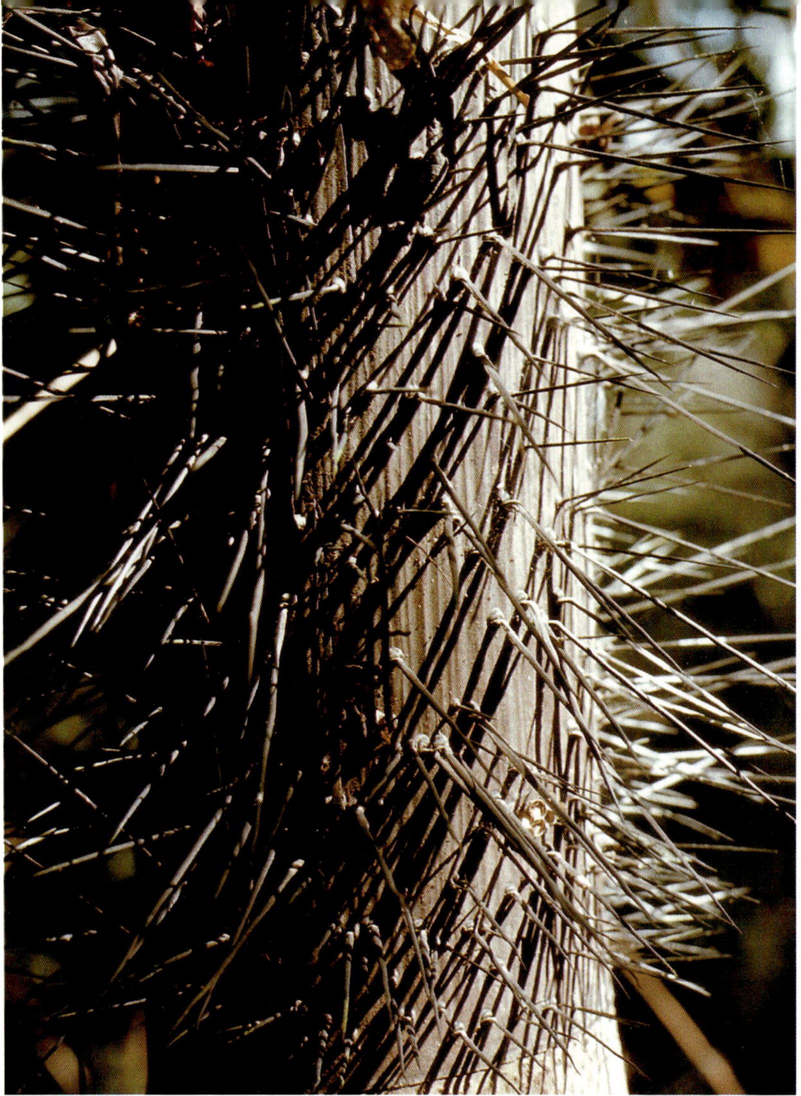

Thorned palm tree.

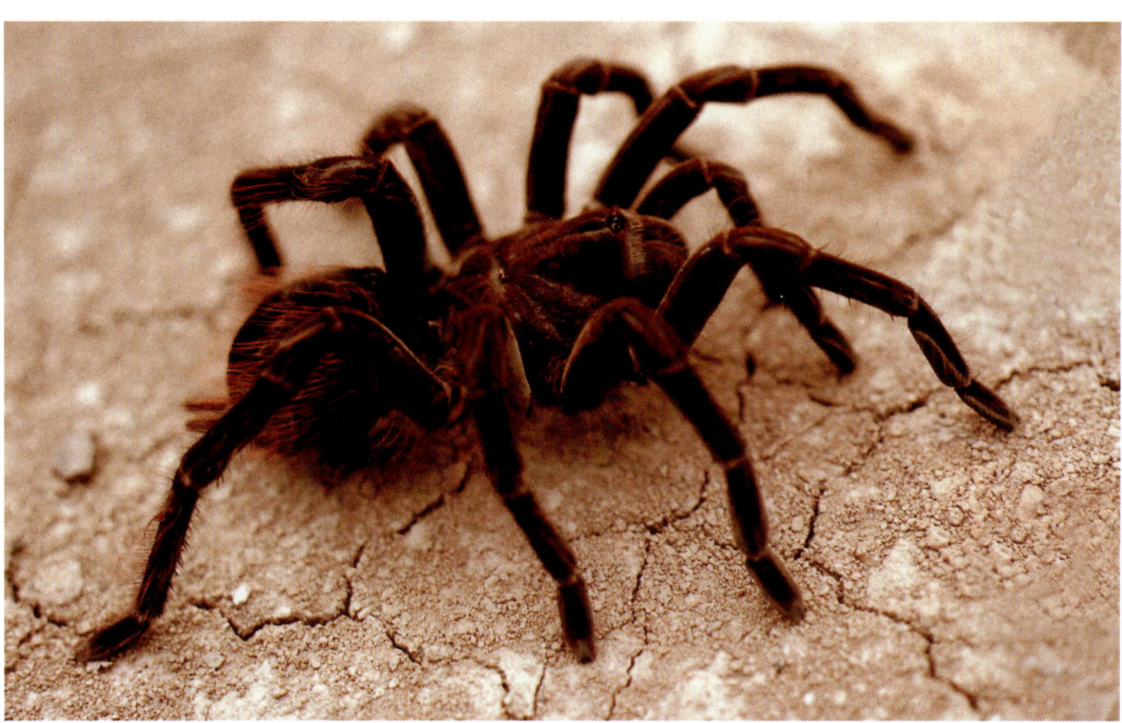

Reaching widths of 20 cm. (8 in.), these enormous hairy, or bird, spiders attack insects, birds, and other small prey.

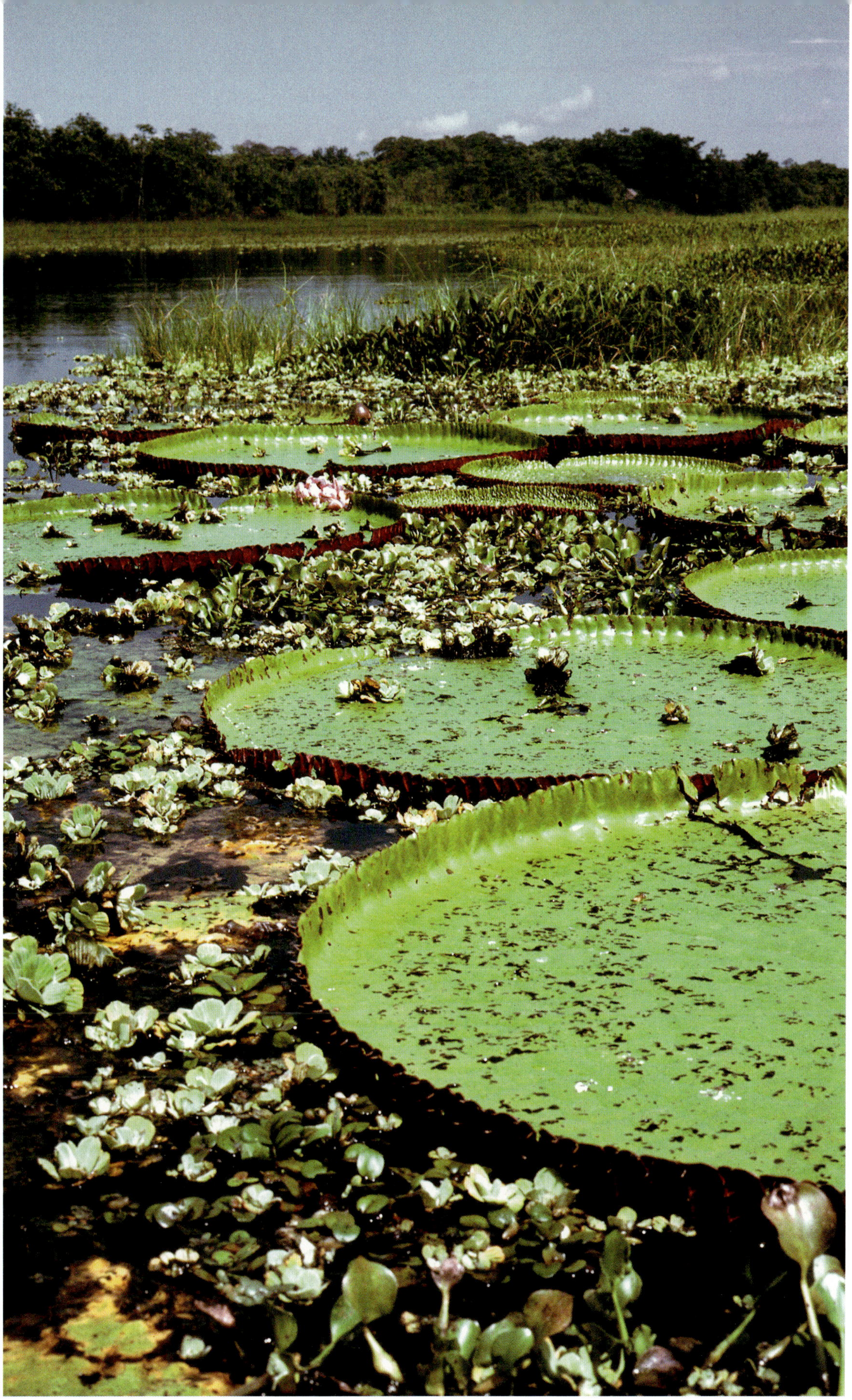

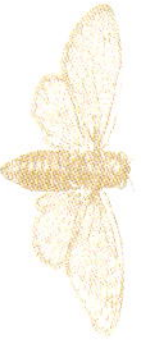

Giant water-lilies *(Victoria regia),* an enormous member of the family Nymphaeaceae from the lagoons of the Amazon.

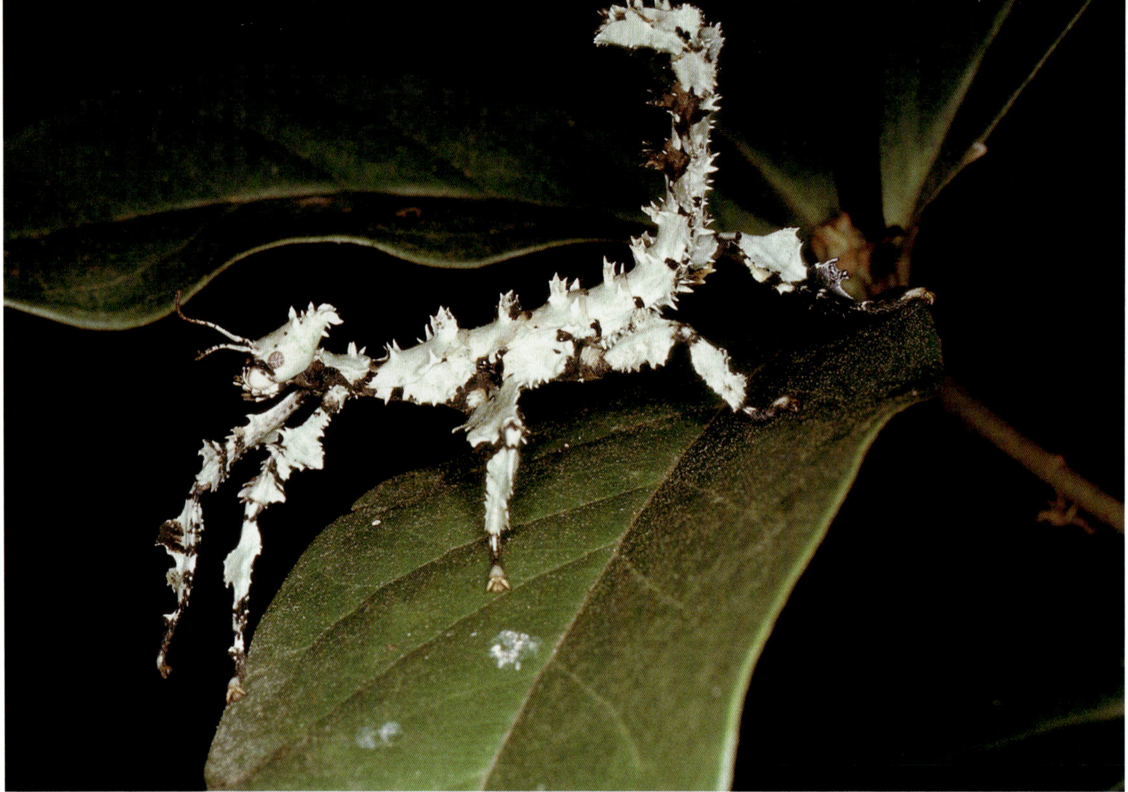

In the tropics nature outdoes itelf, giving rise to extravagant forms, colors, and behavior, together with gigantism. The endless tropical summer may explain this tendency to excess.

Above: This phasmid *(Extatosoma popa)* from Papua New Guinea disappears easily amid the lichens.

The architectural structure of forests thus largely determines insects' spatial distribution. Even though a wide variety of techniques had been used to collect some specimens from the upper forest strata, scientists did not know until recently what an assortment of new creatures awaited them there. This was akin to discovering a previously unknown section of the globe.

Seasonal variation is another important factor affecting distribution: many tropical insect species reproduce throughout the year, whereas the northern varieties appear virtually all at once in early summer.

If we compare the northern and tropical regions, it appears that we see fewer insects where they are most numerous, and we see more insects where there are fewer. Given these different realities, how have the insect populations of each region evolved?

Beginning with the last ice age some ten thousand years ago, the insect populations in northern regions were forced to adapt to harsh climatic conditions, which had no impact on the tropical varieties. In a more optimal environment, the evolution of tropical insects was driven by the pressure of competition within and between species. This may explain why tropical insect species are widely dispersed both spatially and temporally, why many species have very small populations, and finally, why the rainforest contains many rare species.

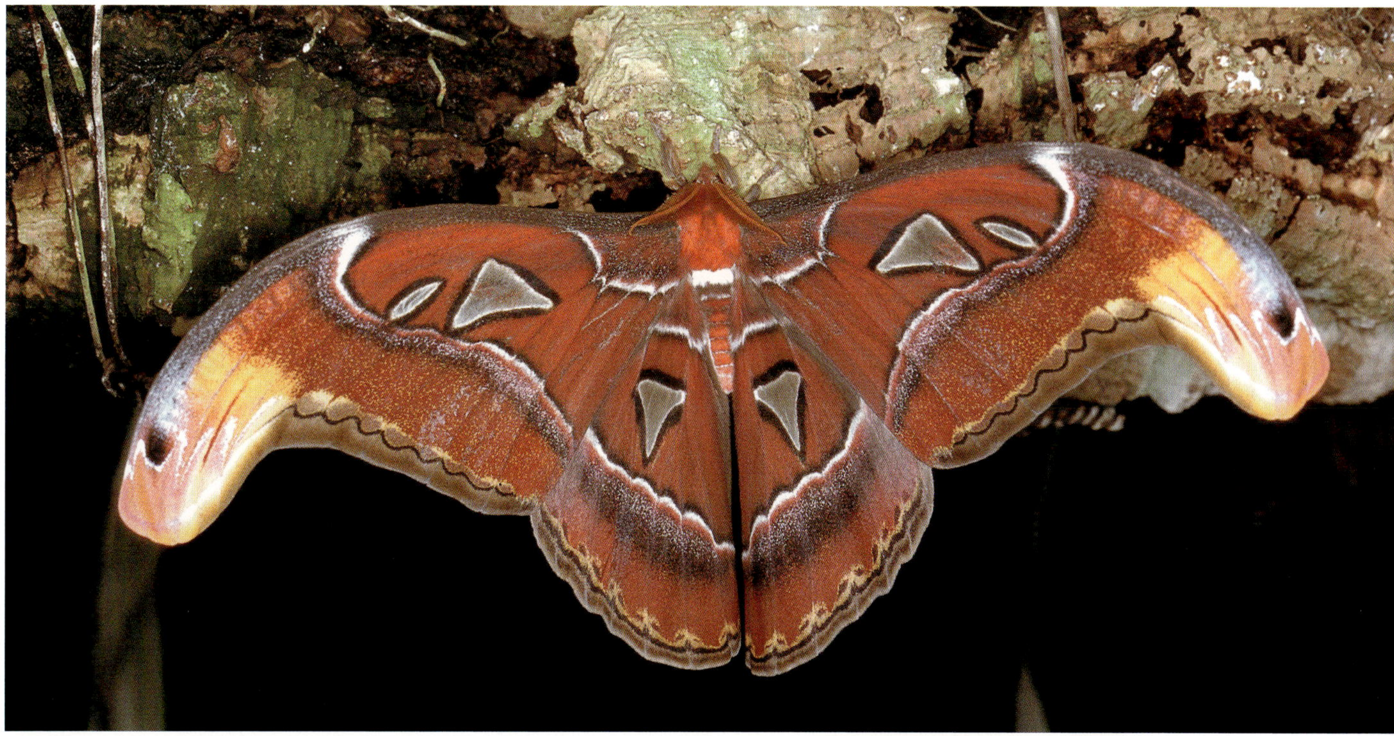

Male atlas moth *(Attacus atlas)*. The genus *Attacus* (family Saturniidae), found throughout Southeast Asia, contains several very large species that attain wingspans of up to 22 cm. (8.5 in.).

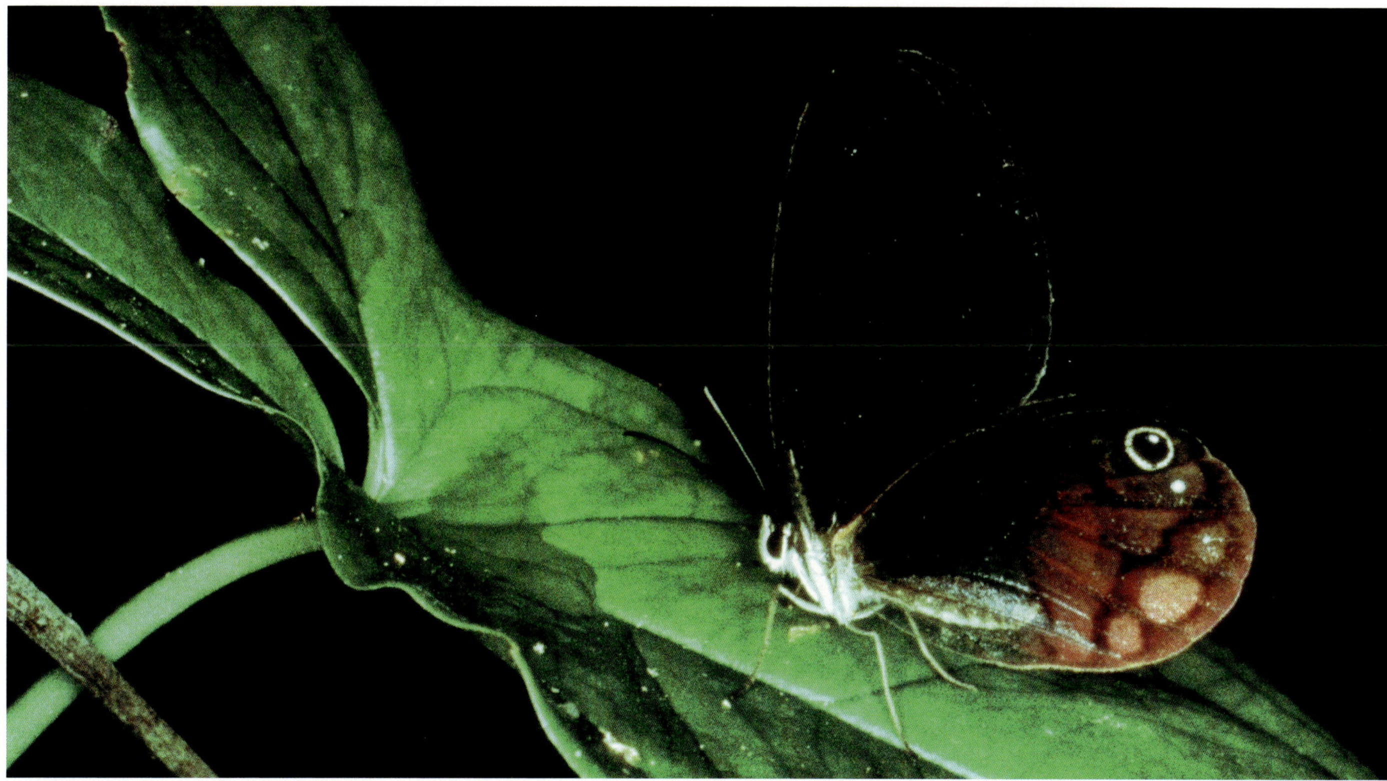

With its semi-transparent wings, this satyrid butterfly *(Cithaerias aurorina)* often confuses its predators. As it flies close to the ground in dim light, its wings provide highly effective protection. This species is widespread in the Amazon.

Insect lovers from the temperate regions may well be disappointed when hunting in the depths of the rainforest. They are likely to see, of all the legendary tropical insects, only those diurnal species—often barely visible—that have adapted to living in relative darkness; the nocturnal varieties are so well camouflaged that very few can be discerned.

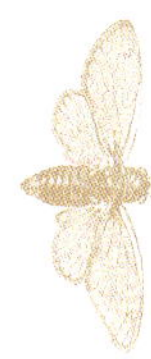

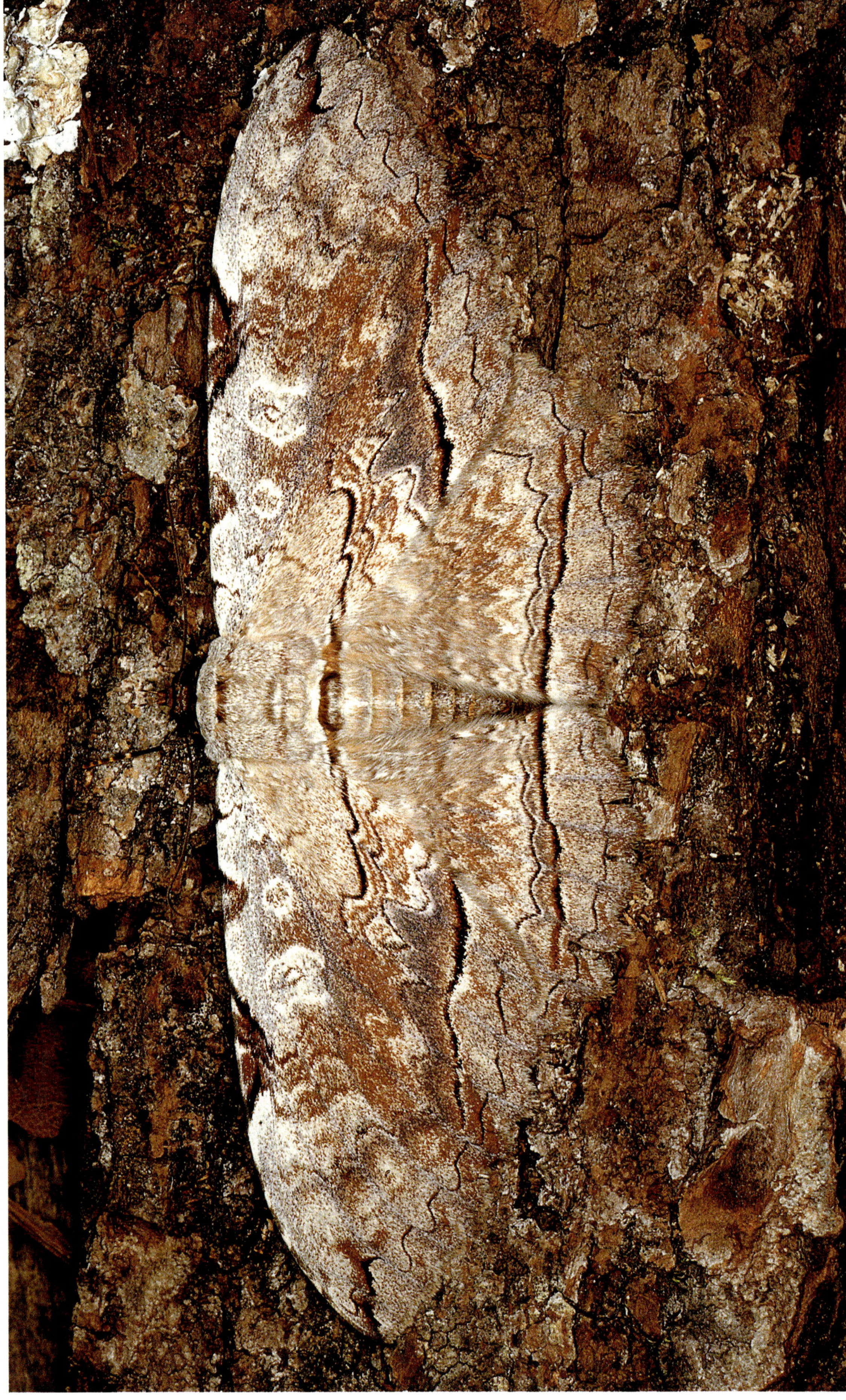

The genus *Thysania* comprises two species, both from tropical America. These large noctuids are experts in the art of camouflage, and their coloration is ideally suited to their dissembling. They also avoid landing right-side up, normally assuming a vertical or oblique posture. *Thysania zenobia* is found across tropical America.

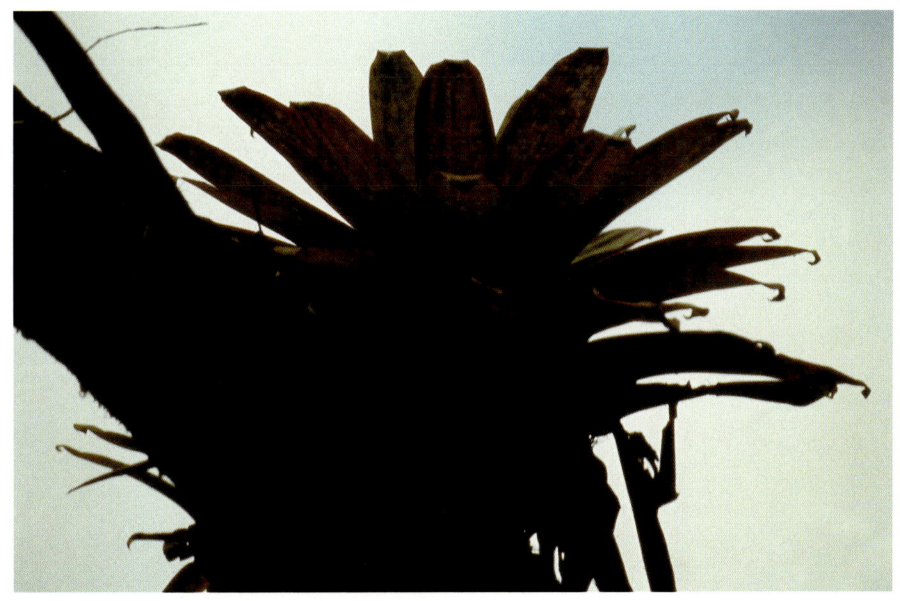

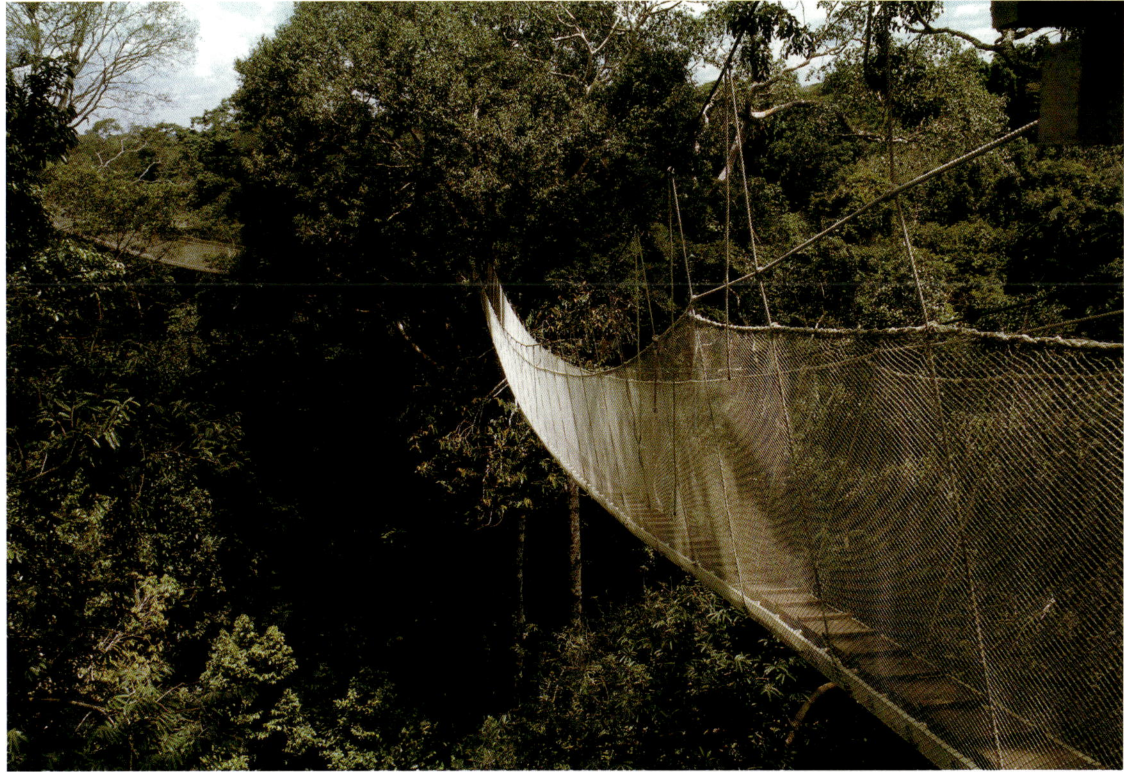

For many years, scientists knew little about the organisms living in the upper reaches of the world's great rain-forests. However, techniques in use during the past forty years have helped reveal the equivalent of a new region of the globe.

Top: Bromeliad plant high in the treetops.

Bottom: Walkway connecting different strata of the Amazonian forest in Peru.

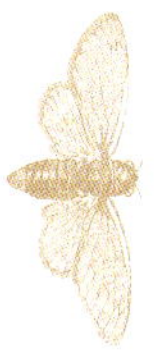

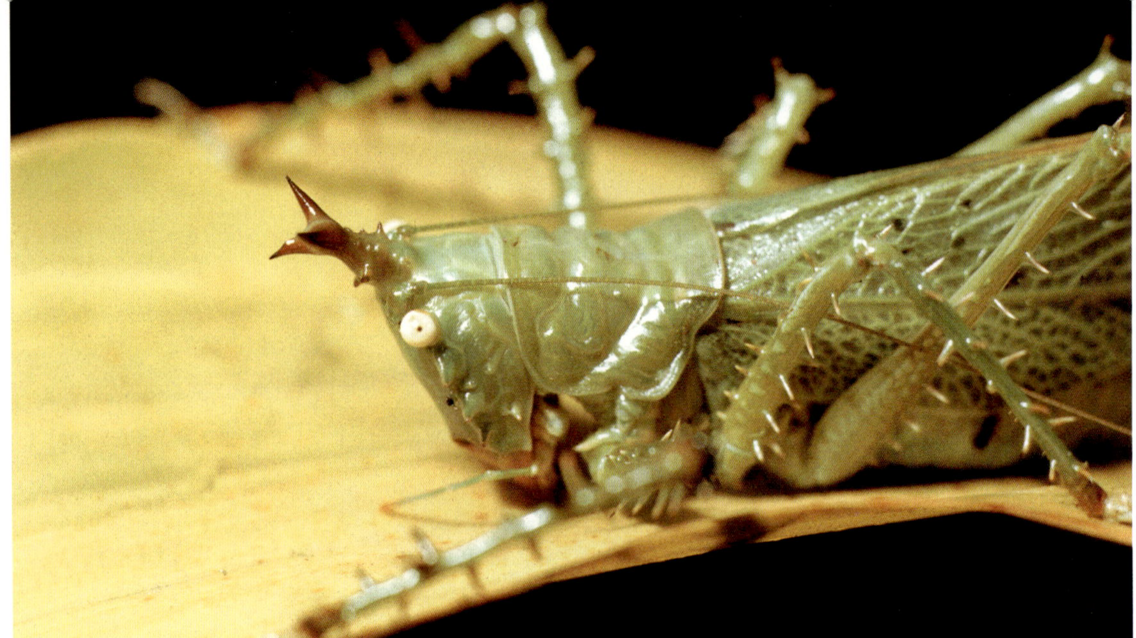

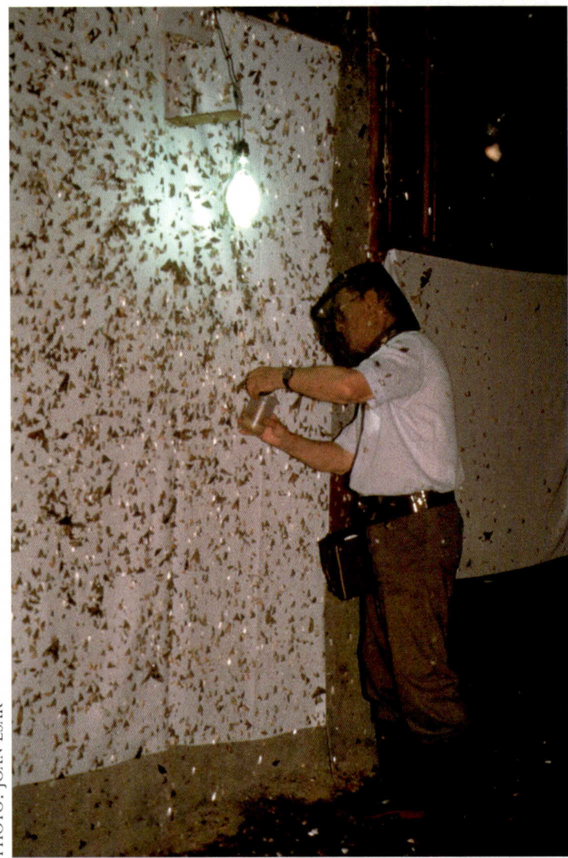

In the rainforest, many species have very small numbers of individuals, and there are a large number of such rare species.

Top: Luck was on my side during my first expedition in Amazonia, when my guide Edgardo caught this extraordinary, rare spiny devil katydid (*Panacanthus cuspidatus,* family Tettigoniidae). During twenty subsequent expeditions in the same region, I have never found another specimen.

Bottom: The author and his light trap.

To compensate for these conditions, insect collectors from northern regions, who are accustomed to finding specimens on a seasonal basis and mostly at eye level, have to rely on light traps when they visit the tropics. Thousands of nocturnal insects from all levels of the forest are attracted by these traps, without which the collector would never see them at all.

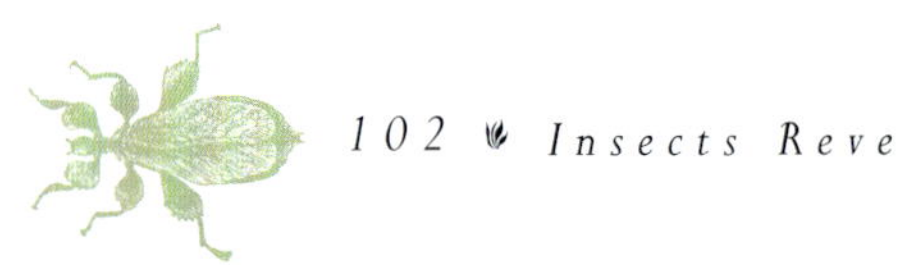

GIGANTISM AND TROPICAL INSECTS

Of all the animal species, the arthropods, particularly the insects, best illustrate the phenomenon of gigantism associated with tropical climates.

Tropical gigantism is typically found among small-sized organisms with relatively short life spans.

The more modest proportions of insect species living in the temperate regions make these gigantic tropical varieties all the more incredible. Even numerous successful collecting trips to the Amazon, Malaysia, New Guinea, and other tropical regions will not satisfy the boundless desire of some collectors for giant species.

Over millions of years, the endless tropical summer has produced such extravagances; indeed, nature in the tropics is constantly pushed to the limit. But the gigantism observed among many tropical insects and other arthropods (such as bird-eating spiders, centipedes, and millipedes) is only one aspect of what appears to be an extraordinary, incalculable, and all-pervading vital force that has amplified insects' forms, colors, textures, and structures, along with their attack and defense behavior, as in no other biotope. It is in the tropical world that nature is most inventive.

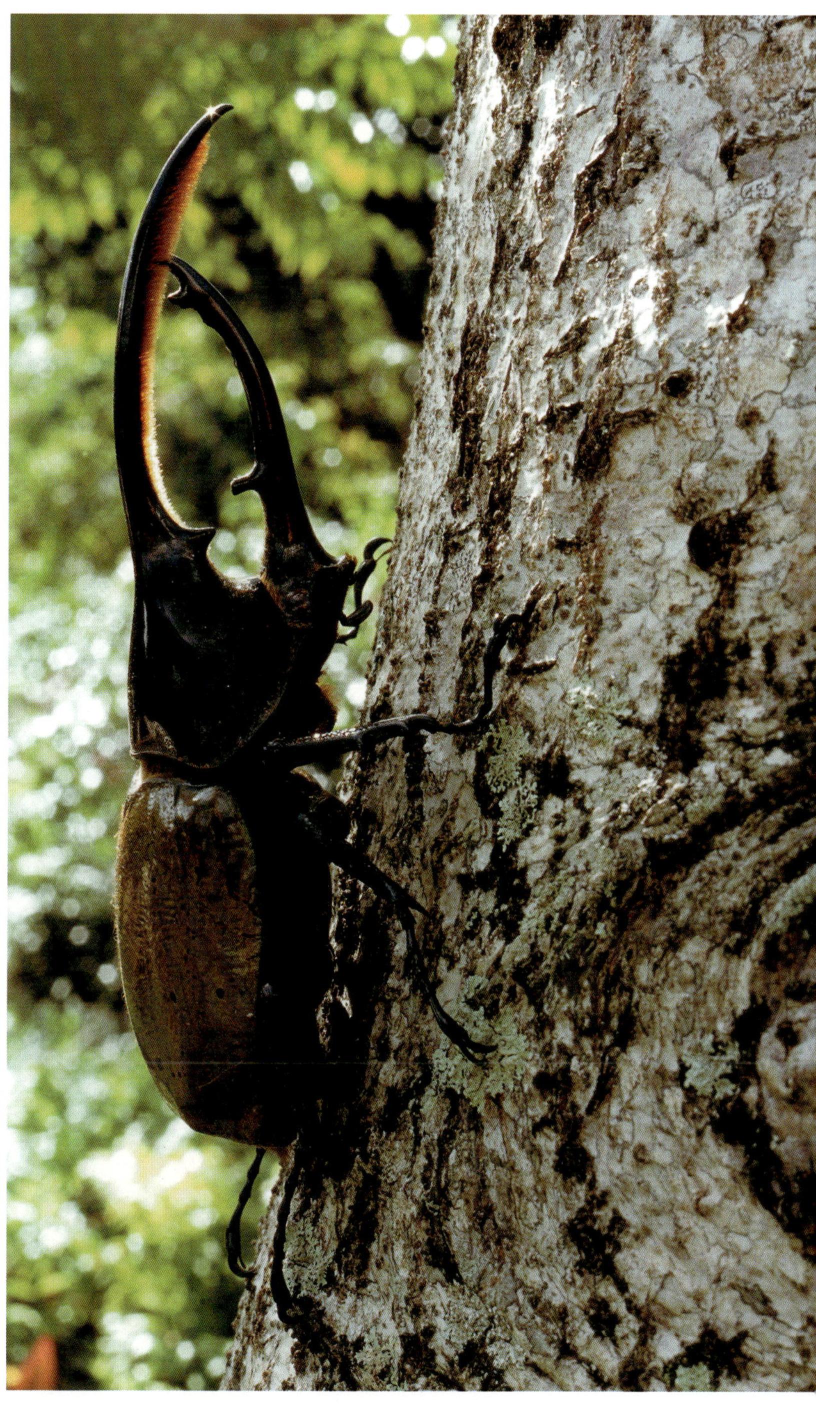

**Goliath, hercules, titan:
Giants of a tiny world**
The male scarab beetles of the Dynastinae subfamily, of which the hercules beetle *(Dynastes hercules)* is a splendid example, have horned outgrowths that are used in combat for females. This beetle (15 cm. [6 in.] long) is a nocturnal species indigenous to Central America, South America, and a few of the islands of the West Indies. The larvae feed on rotting wood and the adults eat fruit.

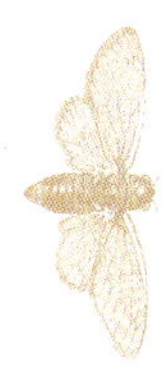

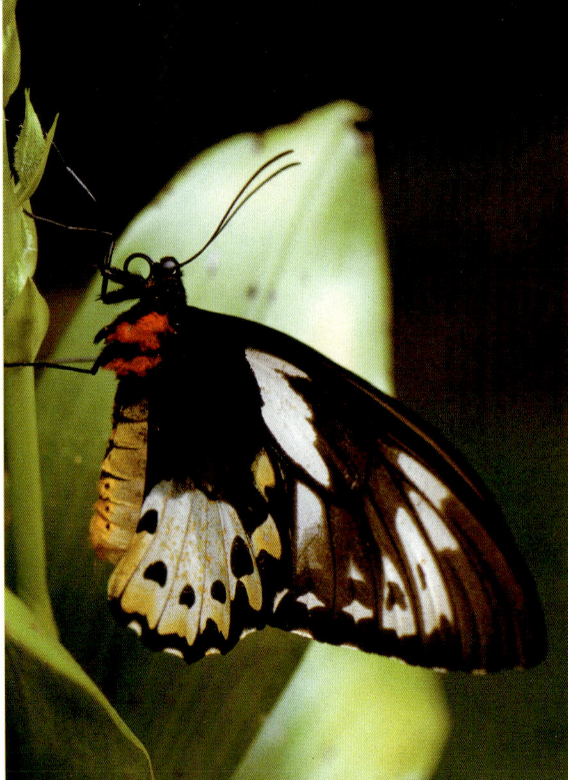

Female poseidon butterfly *(Ornithoptera priamus),* a papilionid from New Guinea. Several subspecies are found on the adjoining islands. Species in the genus *Ornithoptera* are the most regal of the world's butterflies. The wingspan of males is less than that of females (14–17 cm. [5.5–6.5 in.]), and most species are protected by law.

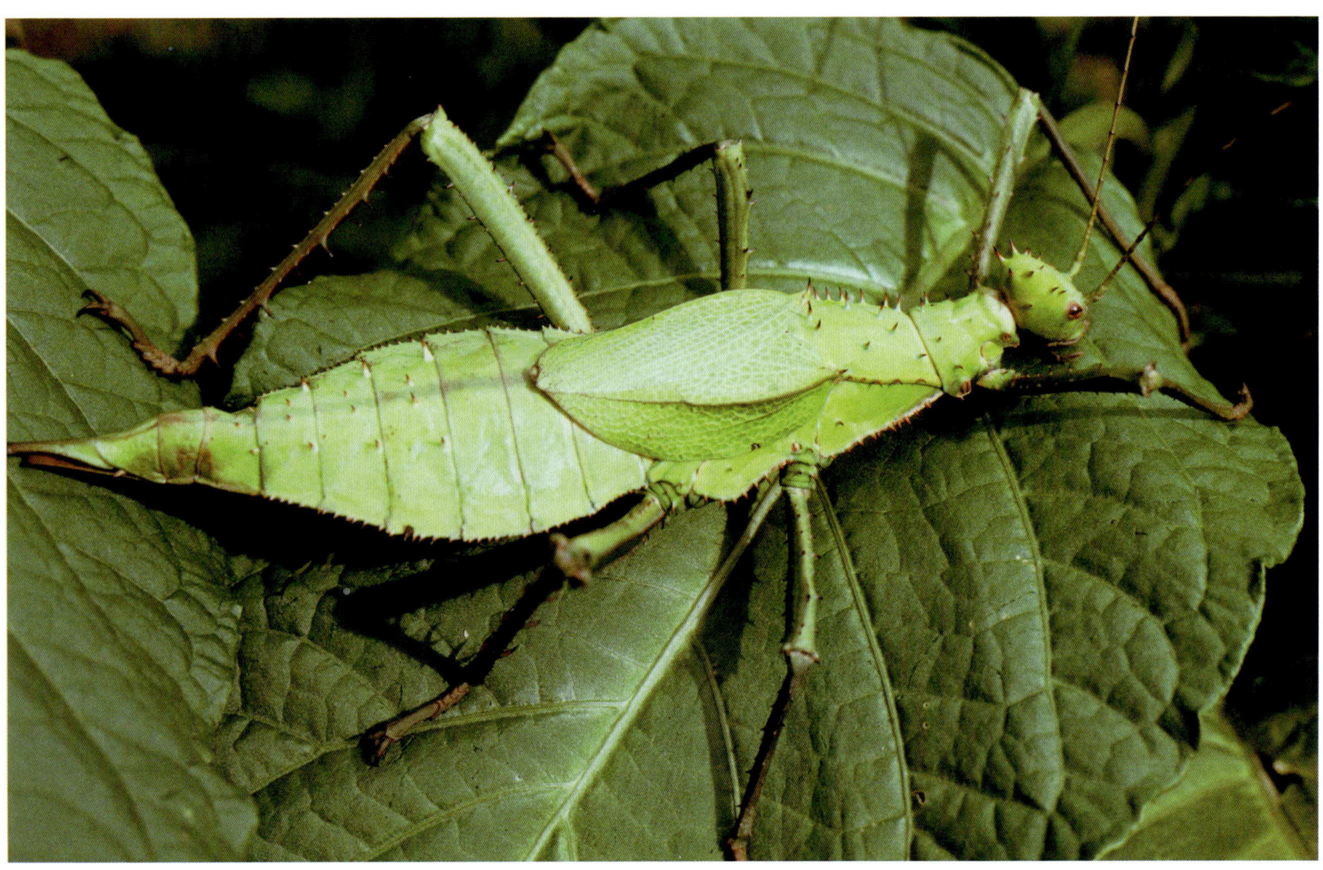

Female jungle, or green, nymph *(Heteropteryx dilatata),* a phasmid from Southeast Asia. The males are brown and are skinny in comparison with the more ample females (14 cm. [5.5 in.]). All phasmids graze on leaves, and some species reach lengths of 22 cm. (8.5 in.), ranking them among the longest insects in the world.

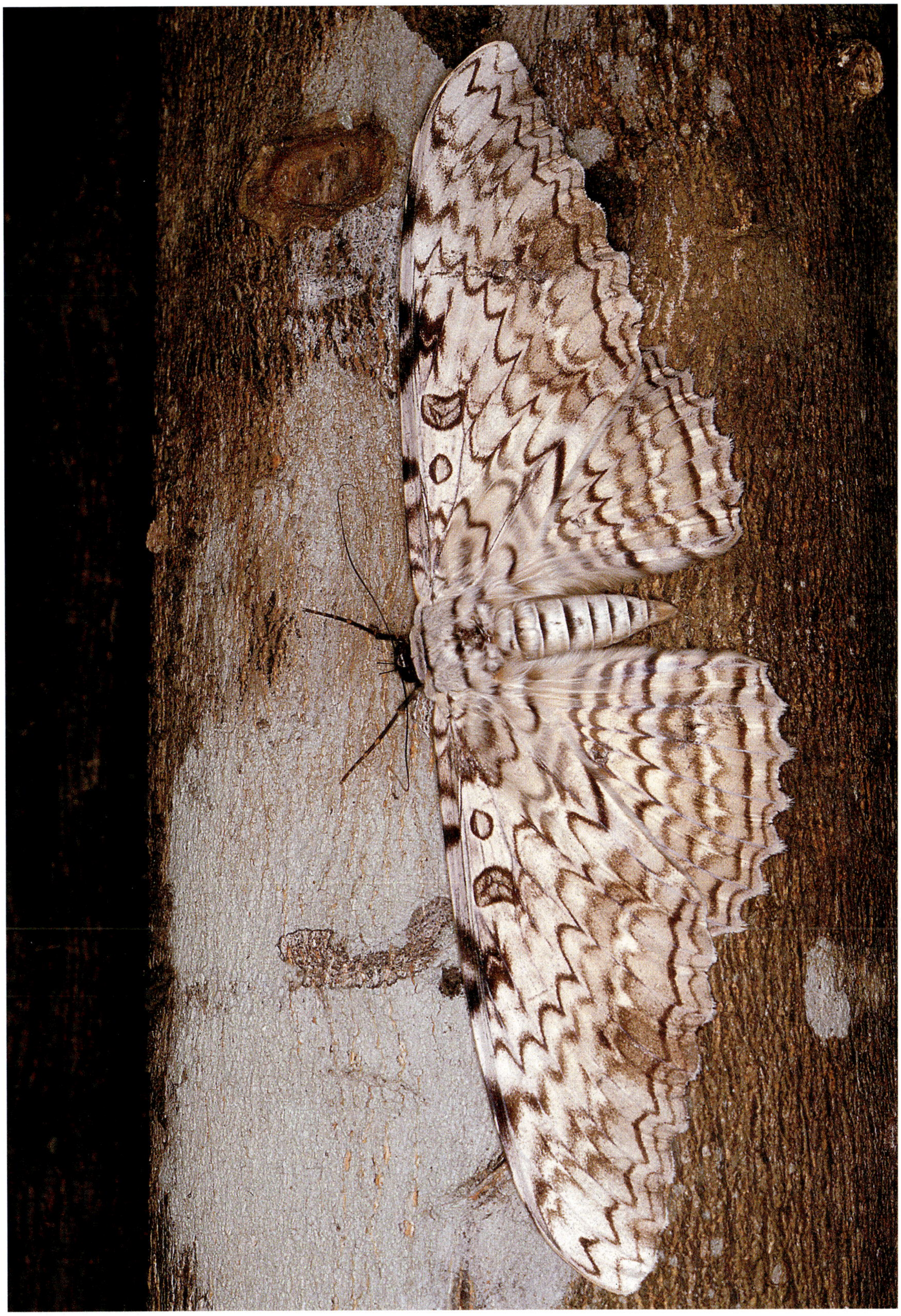

Every tropical region has its giants
This noctuid moth from tropical America, *Thysania agrippina,* holds the Lepidoptera size record, with a wingspan of up to 28 cm. (11 in.). The large *Attacus* moths from Southeast Asia, however, have larger wing surfaces. *Thysania agrippina* displays the same dissimulative tactics as *Thysania zenobia* (see page 100).

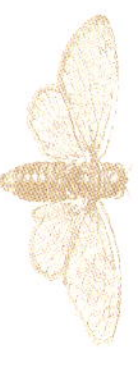

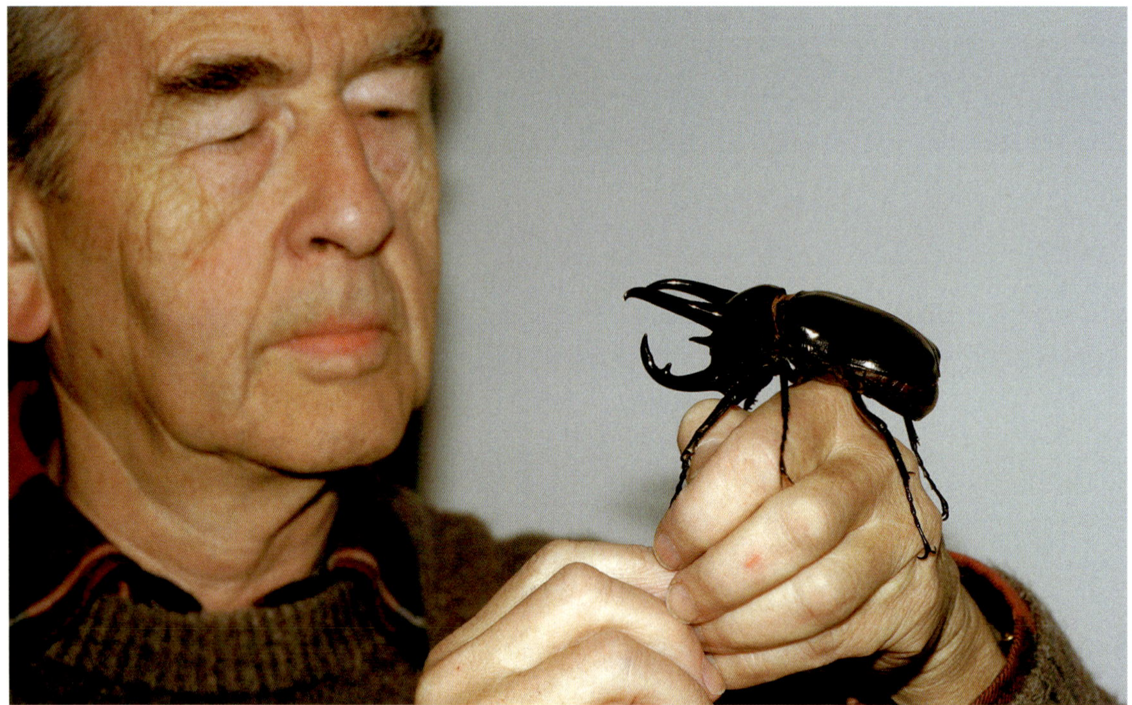

This photograph gives an idea of the relative size of the giant—the giant insect, of course.

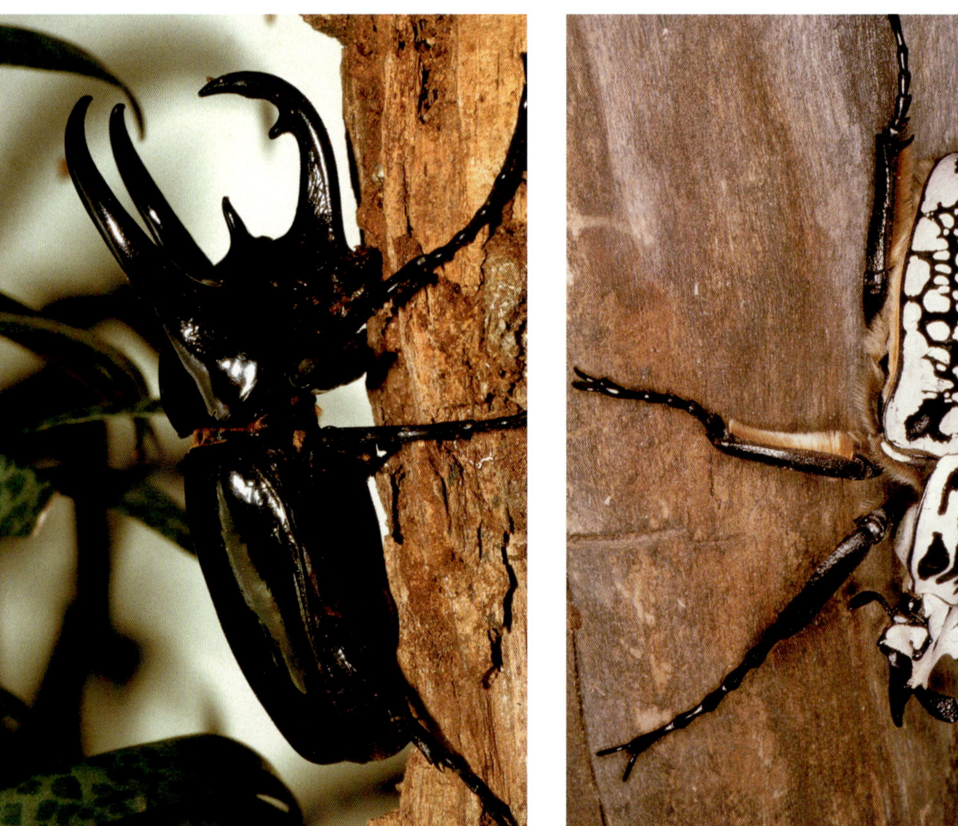

Left: *Chalcosoma caucasus* is one of the world's largest insects. This nocturnal scarab beetle is a member of the subfamily Dynastinae, which is represented by several large-sized species in Southeast Asia and especially in tropical America.

Right: The giant goliath beetle (*Goliathus orientalis,* subfamily Cetoniinae), another colossus of the tropical insect world, is indigenous to equatorial Africa. This beetle and others in the same genus rank among the heaviest insects. Despite their weight, they gather nectar from flowers, sometimes high in the treetops. The goliaths are diurnal and can be up to 10 cm. (4 in.) long.

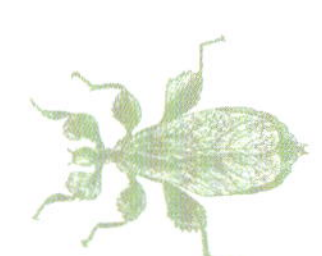

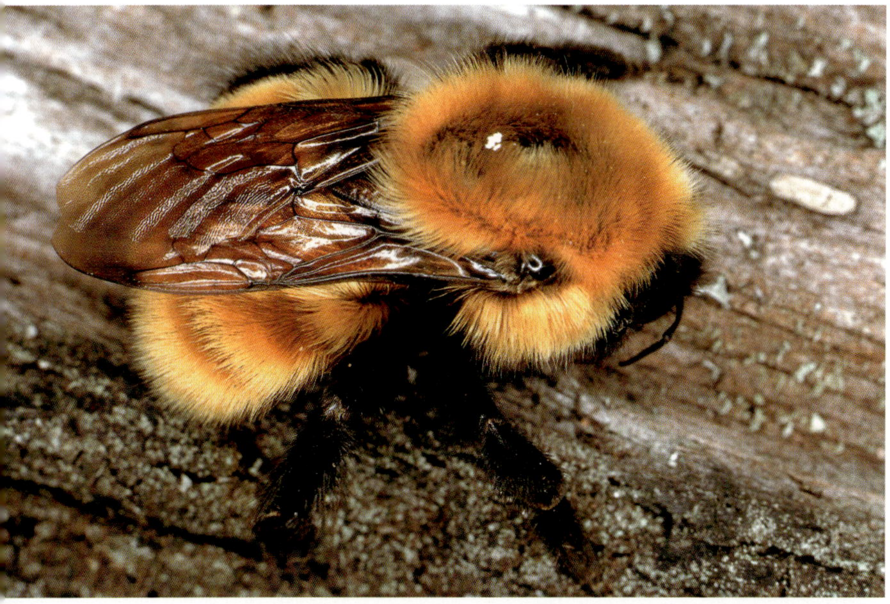

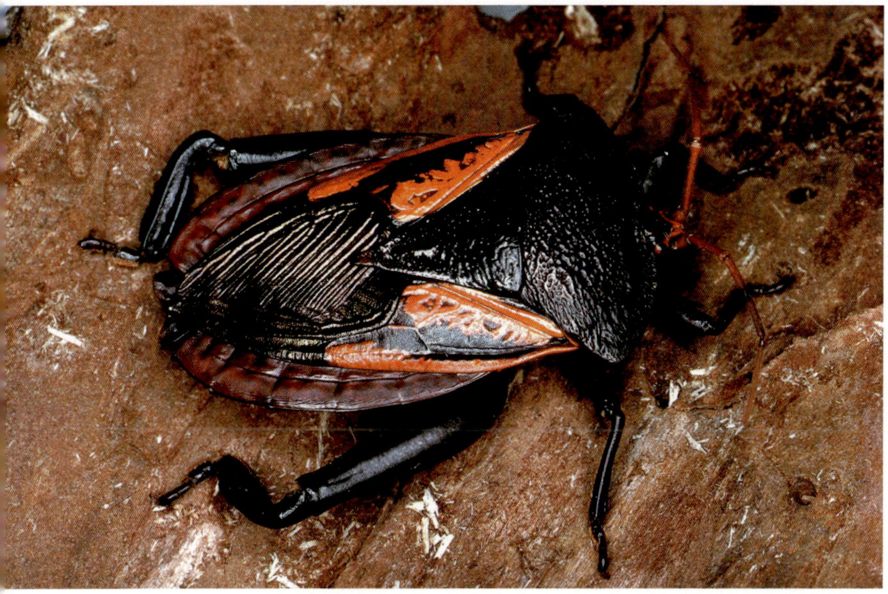

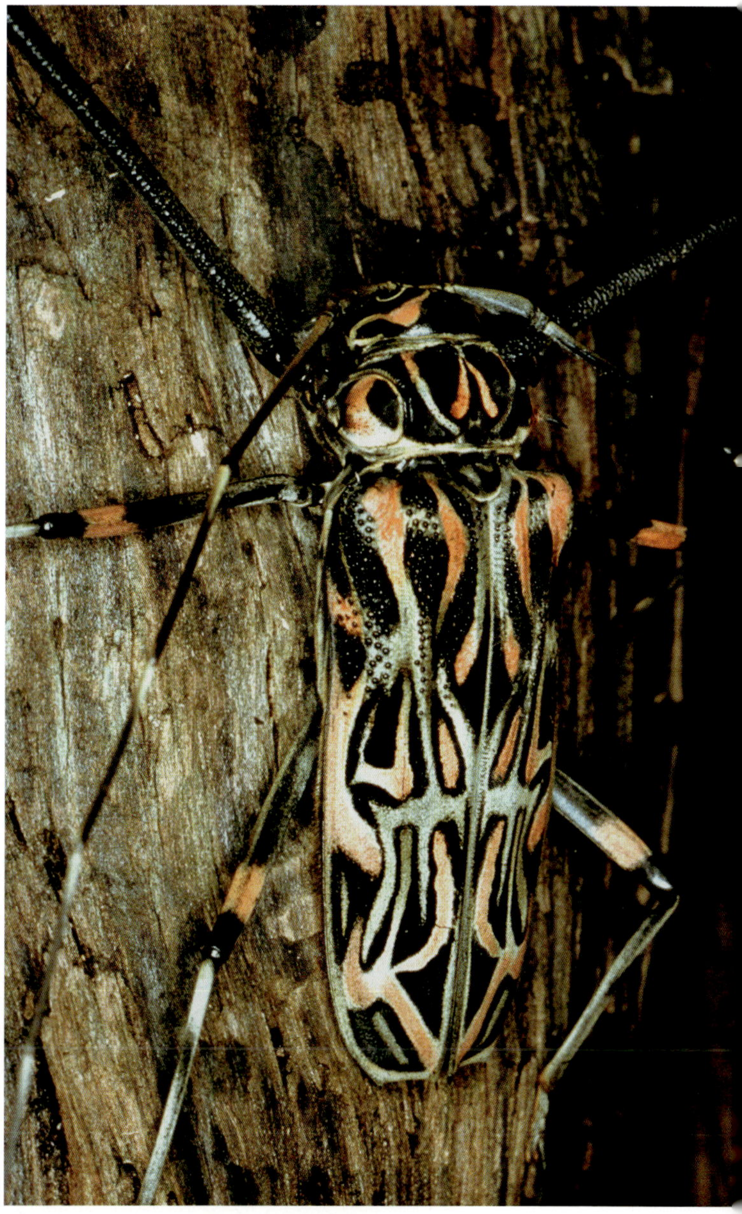

Some insects, although not true giants, earn that title when compared with their more modest-sized relatives.

Top left: Chile's *Megabombus dahlbomi* is at least twice as large as any other bumblebee. The species inhabits the temperate zone of the Southern Hemisphere.

Bottom left: This impressive bug *(Oncomeris flavicornis)* is indigenous to Papua New Guinea (3.4 cm. [1.5 in.]).

Right: This large cerambycid (6.8 cm. [2.5 in.]) from the Amazon basin is the appropriately named harlequin beetle. Its scientific name, *Acrocinus longimanus,* alludes to the extreme length of the males' anterior legs. A member of the subfamily Lamiinae, it is the only member of this group to qualify as a "giant" in a zone where the Prioninae tend to dominate this category.

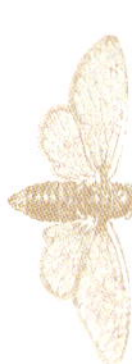

The effects of electricity on insect collecting

As I await the arrival of extraordinary creatures on the sheet or screen of my light trap, I often wonder about the effects of electricity on insect hunting and collecting. What were collections like prior to the advent of electricity, mercury vapor lamps, or ultraviolet tubes, which attract so many nocturnal insects?

The great collections of yesteryear must have been somewhat skewed, since hunting nocturnal insects—much more numerous than diurnal insects—was so difficult. The prevailing view of the insect world was undoubtedly based on diurnal species. Nocturnal insects were able to go about their business undetected and undisturbed while the entomologists slept. Without electricity, capturing a nocturnal insect was a relatively fortuitous event.

The lack of electric lights explains why the beetle *Titanus giganteus* was exceedingly rare even in the largest insect collections until the mid-twentieth century, although Linnaeus had described and named the species two centuries earlier, based on a fragment of one insect. The species is not truly rare, but it does have particular habits; "rare" species are often simply those whose habits are not properly understood.

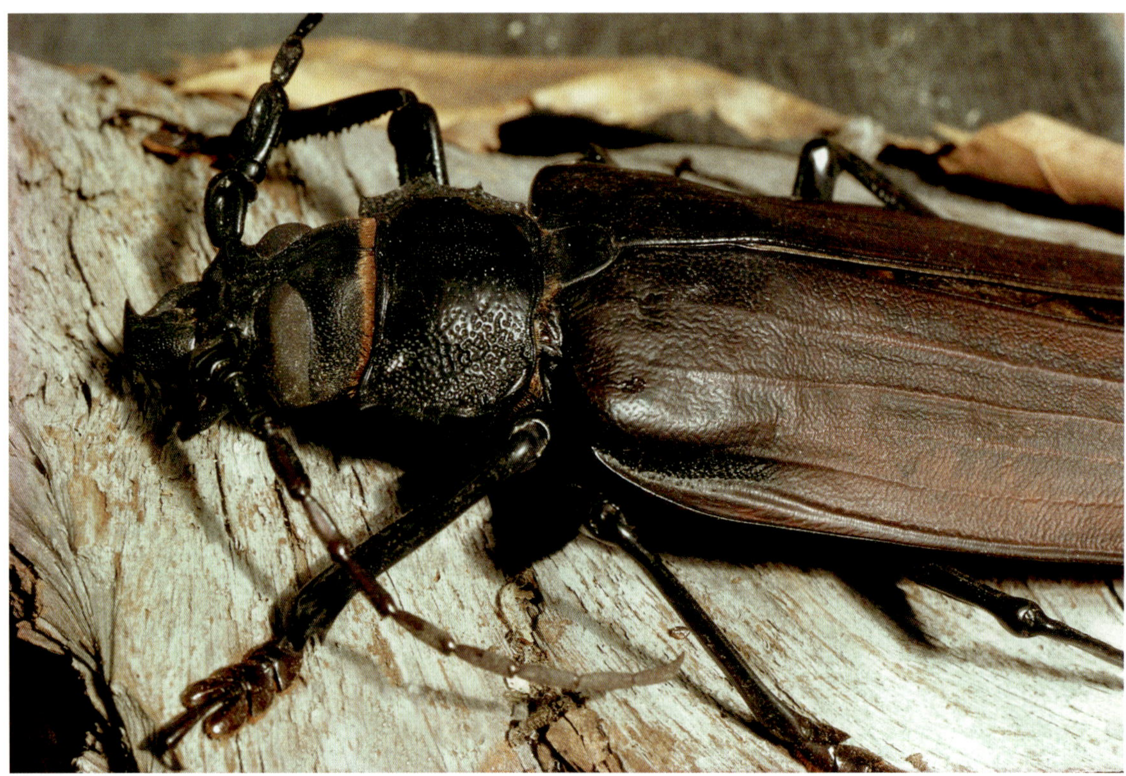

Titanus giganteus is a giant among giants, owing less to its overall bulk than to its length, more than 16 cm. (6 in.). Only males are attracted to light traps, late on moonless nights.

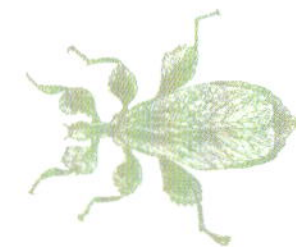

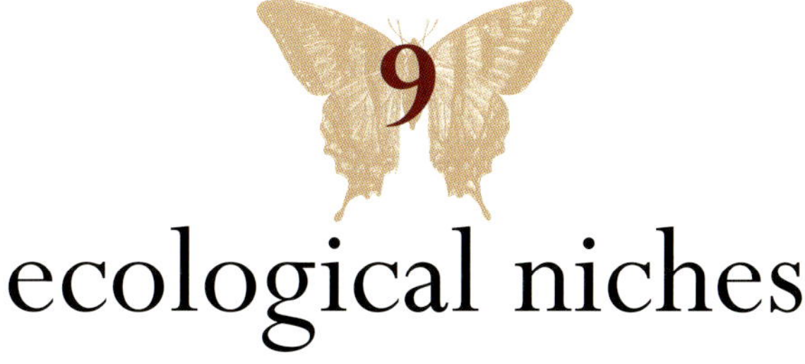

ecological niches

Insects' successful colonization of the planet may be attributed to a number of factors, some of which are external or merely circumstantial. I have already noted that the first insects emerged when the supercontinent of Pangaea offered them a virtually unbroken land mass that fostered their dispersal and that the generally warm and humid climate proved wonderfully hospitable. Moreover, since insects were some of the first organisms to adopt land- or air-based lifestyles, they were able to use a vast array of resources within the environments they occupied.

At all stages of their evolution, insects reaped the benefits provided by these key factors.

However, insects might not have flourished had they not had certain talents or skills—what I like to call their "creativity"—that they have displayed since their very beginnings. This term may seem highly anthropomorphic, but it is the one I prefer. To be sure, all of life's manifestations are creative, and this creativity is not exclusive to insects. But where else is creativity accompanied by such ingenuity and versatility?

It is thanks to insects' creativity, their extraordinary adaptability, that they have managed to take advantage of a wide range of environmental resources in satisfying their specific needs and, at times, in compensating for certain inherent weaknesses.

Insects' diminutive size leaves them vulnerable to any number of predators. But their size also enables them to find thousands of shelters that larger organisms cannot exploit. Another benefit of their size is perhaps even more crucial: small organisms

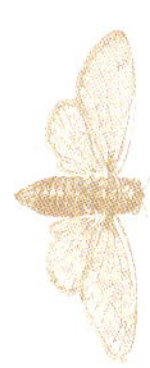

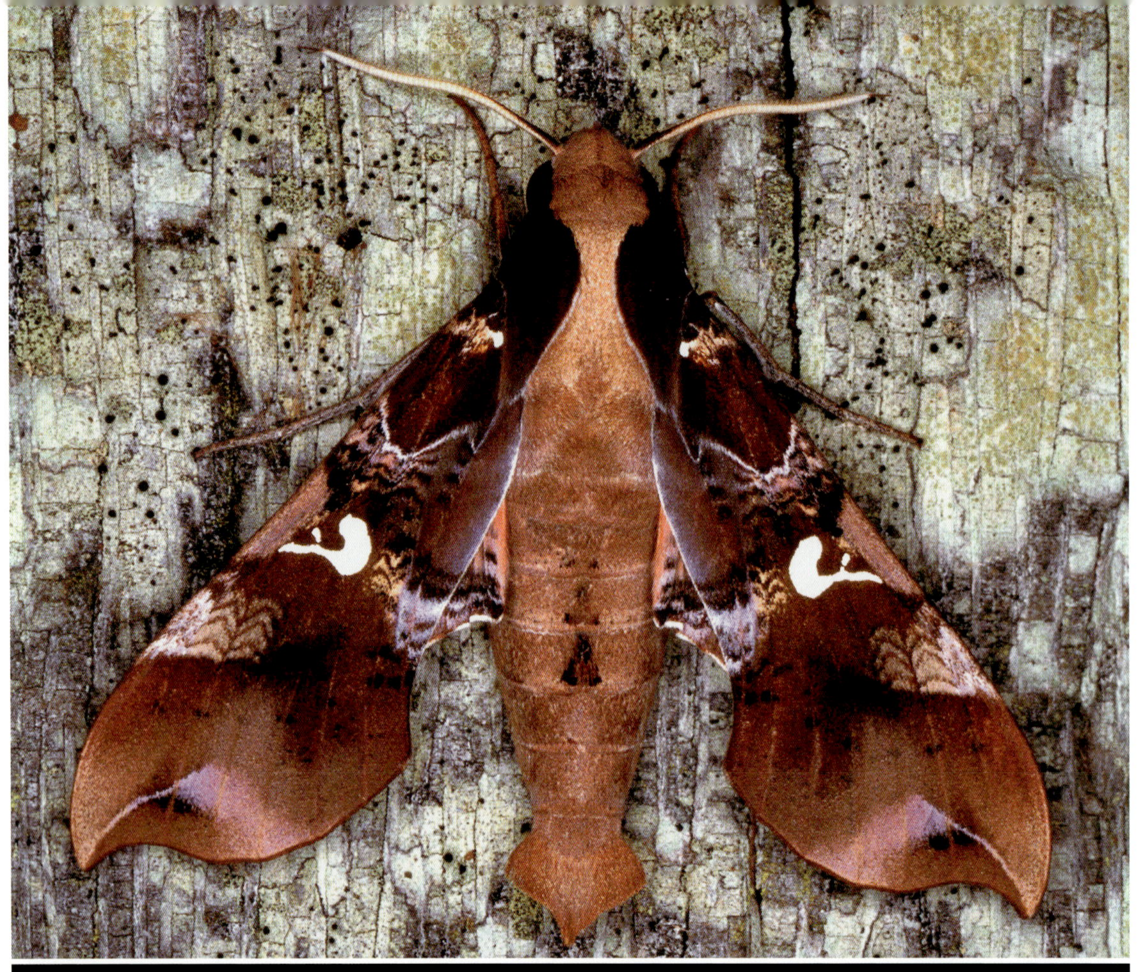

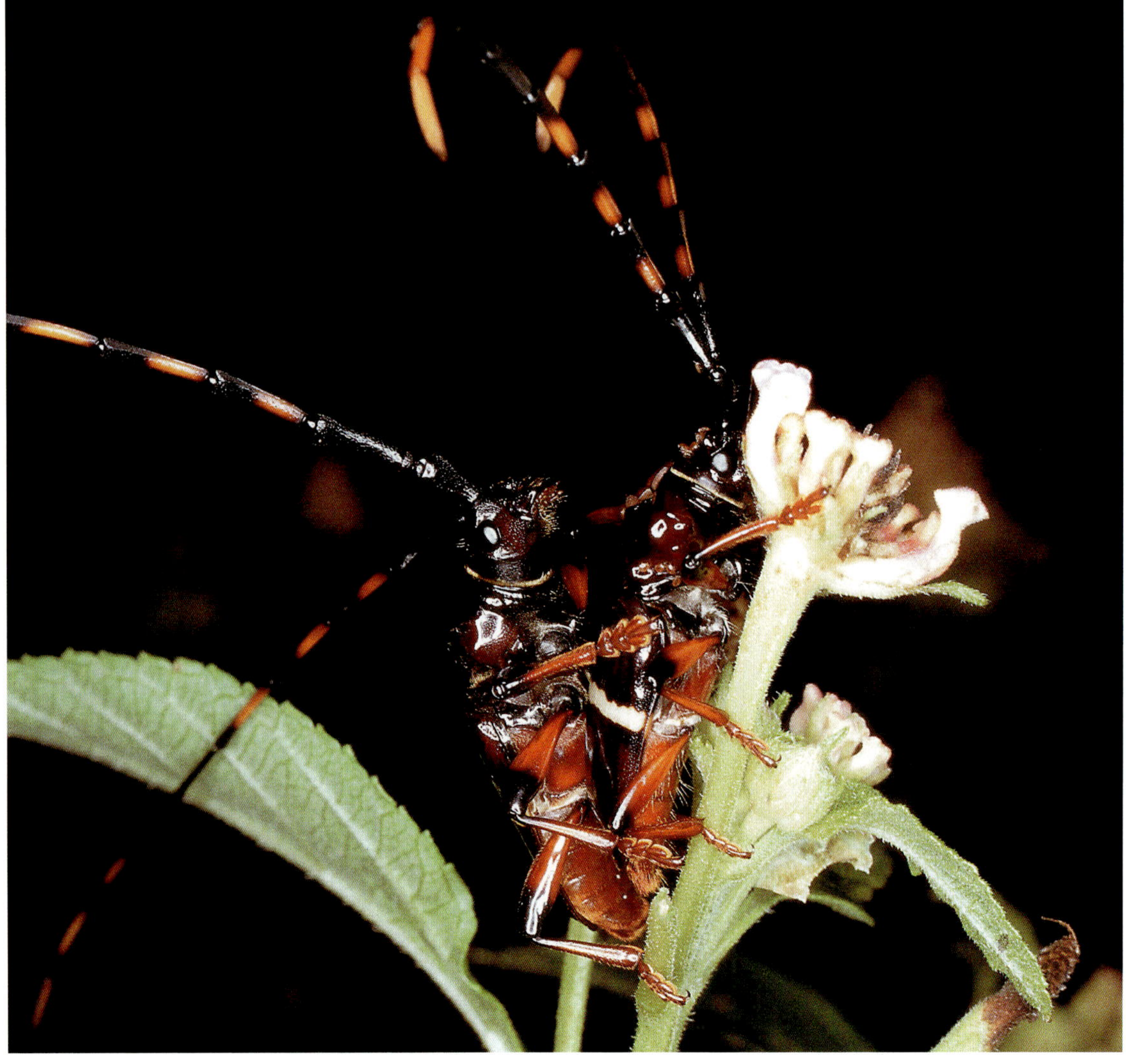

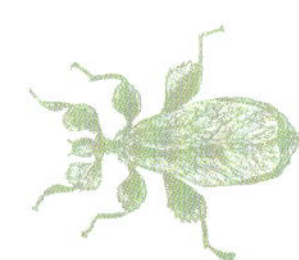

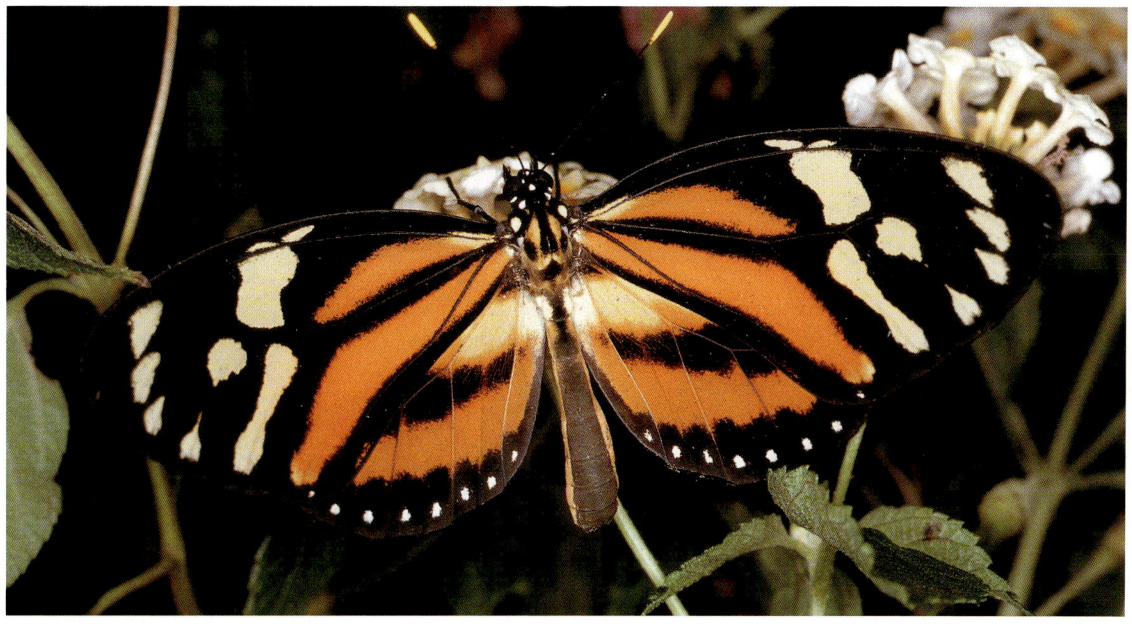

Ecological niches may be understood as abstract structures such as "ways of life" or even "spheres of influence." The niche refers to a species' placement and role within its milieu as it adapts in response to its competitors and predators. This is not so different from the "lifestyles" that city-dwellers devise; after all, Parisians, New Yorkers, and Montrealers normally live in small "villages" within these larger urban centers. Over its short lifespan, an insect fulfills its genetically programmed destiny in relative isolation. It has no awareness of others, unless they are its prey, predators, or sexual partners.

Although these three insects from the Dominican Republic inhabit the same region at the same time, their paths do not cross; their lives are perfectly isolated from each other.

Above: Exclusively neotropical, *Lycorea ceres cleobaea* is a member of the family Danaidae, as is its cousin the monarch butterfly.

Opposite page, top: The parce sphinx moth *(Callionima parce),* a member of the family Sphingidae, is widespread in tropical America.

Opposite page, bottom: Mating of *Trachyderes succinctus*. These beetles (family Cerambycidae) are also common in tropical America.

normally have a high rate of reproduction, which has a bearing on their population density and diversity over the long term. This is particularly evident in tropical climates, where numerous insect species produce several generations per year. Such species are able to adapt more quickly to environmental changes than are species that reproduce only once per year.

During the vicissitudes of the geological ages, the process of natural selection and adaptation resulted in the evolution of large numbers of species, each occupying its own ecological niche.

At first glance, the term "niche" may suggest a hollow space in which an organism might curl up and hide, either to rest or to escape its enemies. However, the ecological niche refers not only to species' habits, but also to many other aspects of insects'

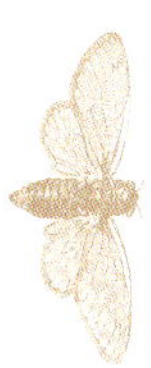

lives. These include how species use the space available to them and how they develop over time, in conjunction with the other species that share a particular ecological community, or biocenosis.

It is easy to confuse species' ecological niches with the territories they occupy. Territory is only one component of an ecological niche, which is a broader and more abstract concept. Although the ecological niche does refer to a structure, it is a structure in the sense of the terms "way of life" or "sphere of influence."

The ecological niche designates the placement of a species within an environment, together with the role it plays therein. Seen from inside, a niche is determined by the pressures exerted by a species on its environment as it exploits all available resources within the limits imposed by neighboring species. An ecological niche thus forms as a result of pressures and compromises with competing species. Two species living in close proximity and seeking to exploit the same environmental resources could not coexist; one species would dominate and exclude the other. Through the play of pressure and compromise, the weaker species would modify its behavior to adjust to the situation; however, it would no longer be able to use the same resources as those used by the dominant species.

Ecological niches are thus founded on a fortunate compromise.

If conditions are stable, each species achieves ecological isolation within a well-defined ecological niche; any energies that might be expended on strenuous inter-species competition are thus reduced. In such an environment, each species conforms to the limits imposed by its niche, never straying across the boundary of a neighboring niche, in much the same way as the well-defined segments of an orange reflect a balance of opposing forces.

Some insect species occupy extremely restricted ecological niches. Consider, for example, those parasites whose only connection to their external environment is the body, nest, or lair of a bird or mammal. One tiny beetle, for example, is found only in beaver fur and is transmitted from a mother beaver to its offspring. These beetles have no choice but to go underwater or on land according to the activities of their host. Another specialized ectoparasite spends its entire life buried in a sheep's fleece and travels only a fraction of an inch to draw its ration of blood or to reproduce.

Not all creatures lead such obscure and restricted lives. Some powerful species flourish brilliantly within their biotopes and seem to carve out starring roles. For example, monarch butterflies *(Danaus plexippus)* adapt to an extraordinary variety of conditions around the world and make their homes equally well in the Amazon, New Guinea, or Canada.

In summary, an ecological niche is determined primarily by the quality and quantity of resources offered by a habitat. However, the niche is actually defined by the way a species uses the resources according to its needs and skills, within the limits set by its competitors.

If we imagine the niche as having a physical form, the boundaries would be determined by the pressure a species exerts on the environment through the role it plays.

One can pose many questions regarding this role, which resembles a human occupation or profession, a few of which I list here. From which type of habitat does a species come: the boreal (northern) forest, the humid tropical forest, the savanna, or the desert? From which of the world's many zones? What type of space does the species occupy within that habitat? In which season does the species appear? In the case of seasonal organisms with short life spans, such as insects, how many generations are there each year? Is the species diurnal or nocturnal? Solitary, gregarious, or social? Carnivorous, phytophagous, or coprophagous? Which species compete with it? Which prey on it? We could add a thousand other questions to this unsystematic list; together, the answers would enable us to determine the outlines of the invisible structure known as an ecological niche.

Computers, which can be used to analyze the complex networks of such interacting systems, may help us to understand niches in all their structural subtlety, thus making them as real for us as they are for their inhabitants.

A flourishing ecosystem is thus composed of harmonious niches, all interlocking to form a balanced architectural structure.

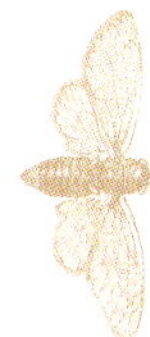

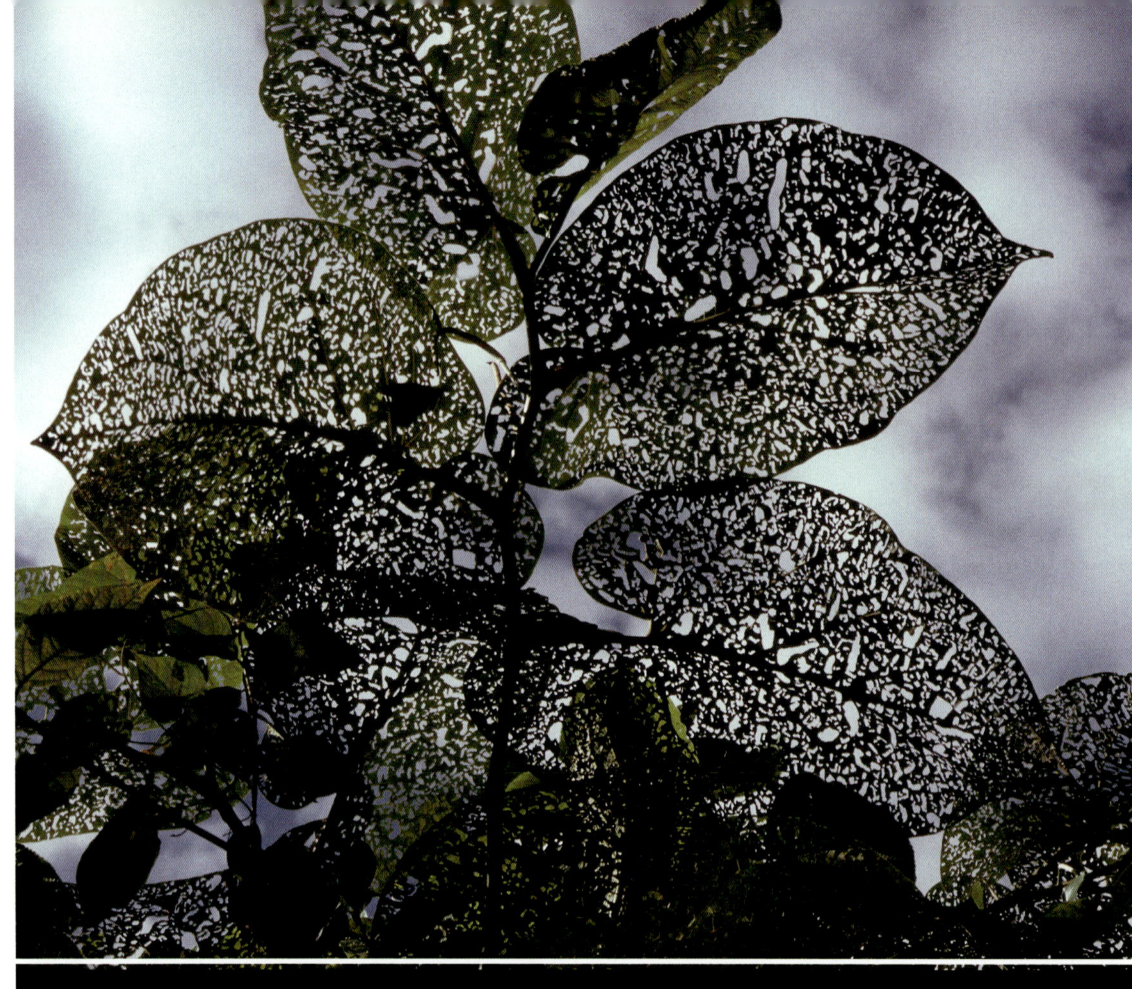

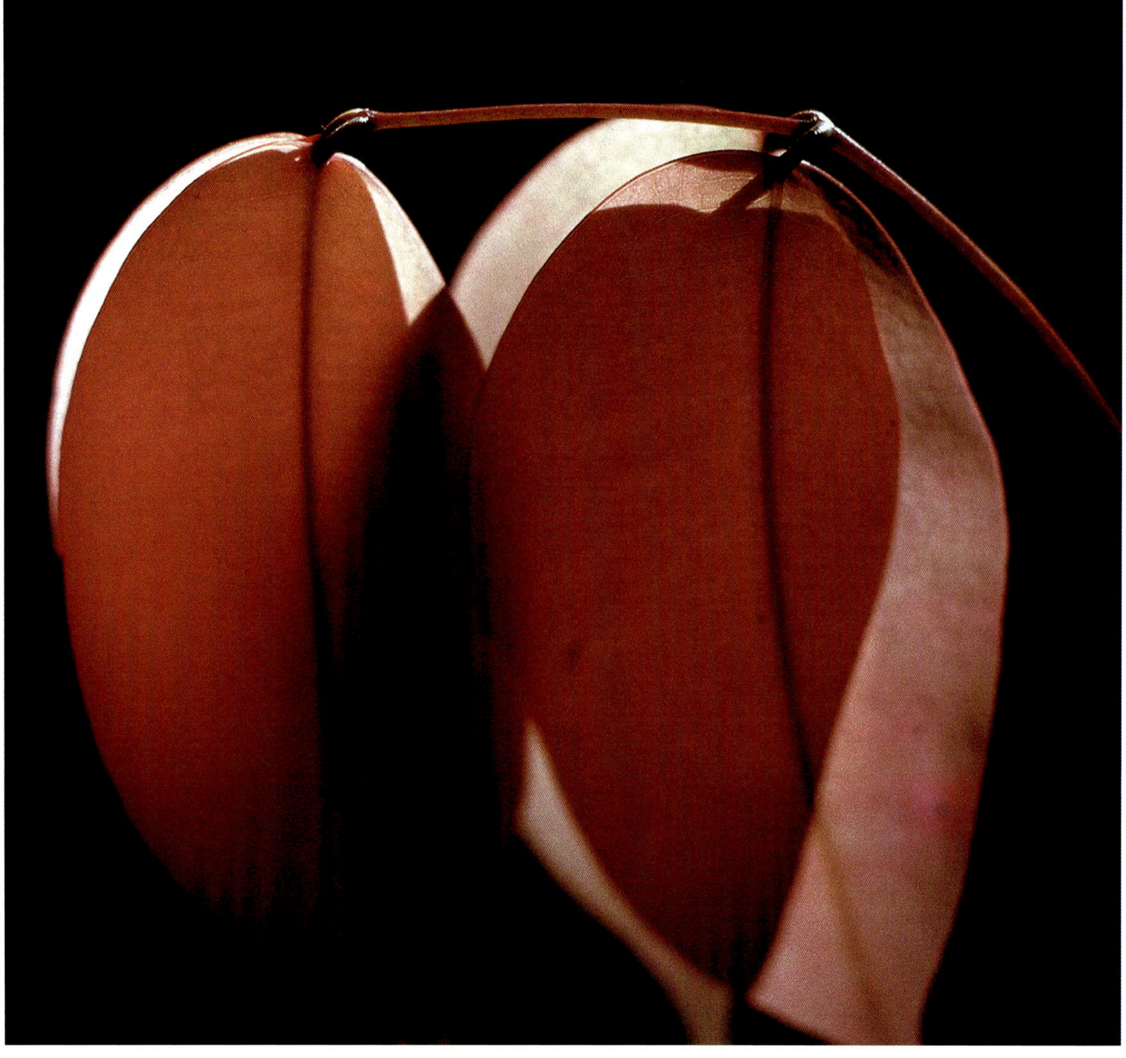

10

insects and plants

In chapter 4, I described the idyllic and interdependent relationship between flowering plants and their pollinators, the nectar-gathering insects. This relationship may indeed be very beautiful: butterflies zip from flower to flower—an abiding poetic image—drinking the sugary nectar and covering themselves with pollen, thereby fertilizing other flowers. But bear in mind that these same butterflies were at one time caterpillars, and there is nothing at all idyllic about caterpillars' passion for plants.

Bees are perhaps more chivalrous in their courtship since they never take from plants anything but what the plants' flowers owe them for pollinating services rendered.

Plants and insects have formed other types of rewarding relationships which, although based on different types of exchanges, are nonetheless admirable for their mutuality. For example, the myrmecophiles (ant-friendly plants) have close links with ant colonies. A myrmecophilous plant provides nourishment for its resident ants; these in turn defend the plant against any insect or other animal that threatens it, and even against other plants that might seek to twine themselves around it. In tropical America, certain varieties of acacia are equipped with enormous thorns resembling bull's horns; with their armies of tiny guardian ants, which live in the hollow thorns, the trees become impregnable fortresses.

Most relationships between plants and insects, aside from the amicable ones just reviewed, might be described as open warfare. Hordes of herbivorous insects attack plants the world over; in fact, their mouth parts are so varied and their appetites so voracious that we are lucky that any greenery is left on the planet. Fortunately, there

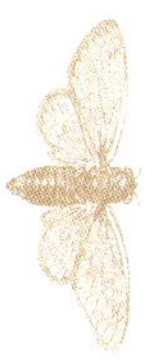

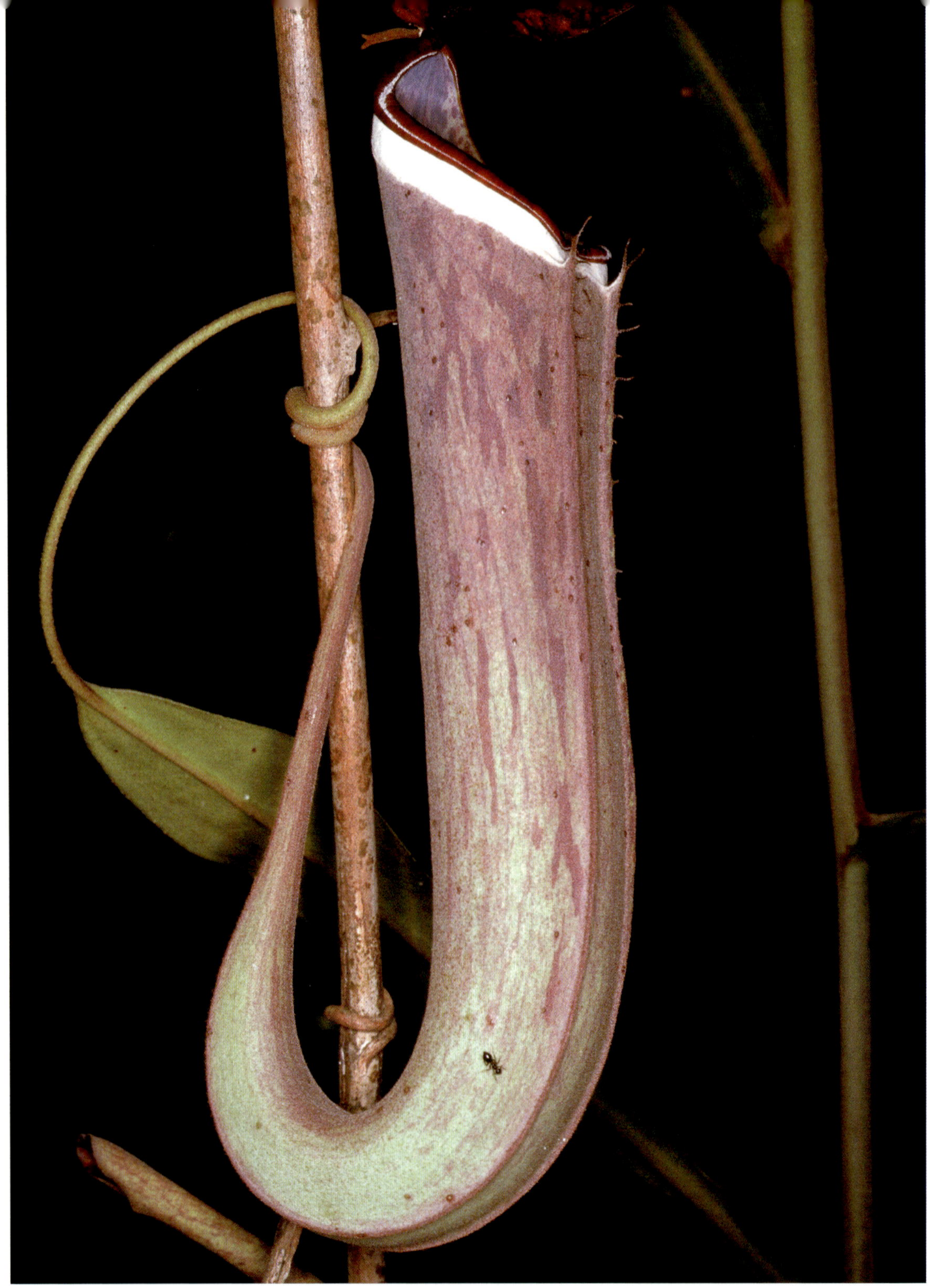

Most insects are either phytophagous or xylophagous during at least one stage of their existence. Few plants escape the ravages of either larvae or adult insects; every plant contains an essential resource that some insect will eventually claim. Plants and insects have evolved together. Insect attacks have prompted plants to take defensive action, even though they may seem completely powerless.

Above: A role reversal: Nepenthes are insectivorous plants from Southeast Asia (family Sarraceniaceae, or pitcher plants). The urn-shaped structure contains juices that digest any victim that falls inside—a plant's silent vengeance.

Preceding left page, top: This Costa Rican plant, whose leaves were almost totally devoured by chrysomelid beetles, still manages to survive because the beetles did not touch the leaf veins.

Preceding left page, bottom: More so in the tropics than in other regions, many young shoots display colors that do not appeal to herbivores. This merely serves as a reprieve, since once they turn green, they will fall victim to numerous well-sharpened mandibles.

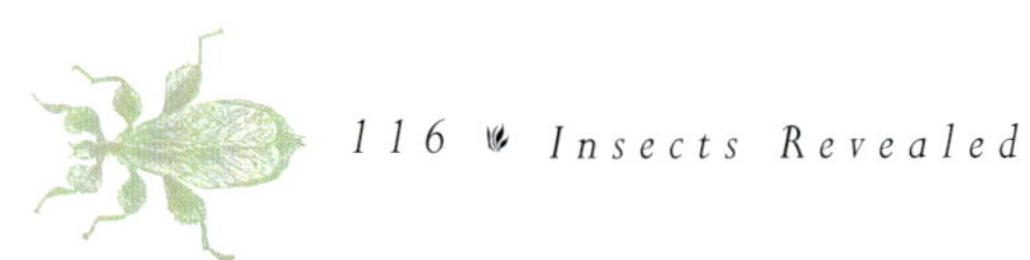

has always existed an equilibrium between plants, which are often mistakenly viewed as passive and defenseless victims of their attackers' sure-fire techniques of destruction, and insects, each of which has its own particular specialty. Some insects chew or grind, others make incisions, suck, or weaken some part of a plant, and no plant can escape their predations. There will always be an insect ready to stake a claim to its own portion of a plant, be it terrestrial, aquatic, herbaceous, or woody. Every plant has its tormenters, which are often highly specialized since plants rarely serve as hosts to a single species. Different insects go after plants during different seasons and thus feed on different parts of the plant's structure, including the buds, the flowers, and the leaves. In the interim, the plant's roots and stem or bark come under attack as well.

As part of an ancient dynamic, insect-plant coexistence opens a window into complex interrelationships that science is only just beginning to understand. Under attack by insects, plants have not submitted and sacrificed themselves as passively as outward appearances might lead us to believe.

Visible defense systems, such as the thorns and spines found on cacti, certain species of palm tree, and acacias, may serve to discourage attacks by large herbivores. However, they have little deterrent effect on insects, which wander among them as we might stroll through a forest.

Plants' anti-insect defenses also include chemical warfare. Indeed, plants made use of pesticides millions of years before humans ever did.

In the words of Adrian Forsyth and Ken Miyata, "More than 10,000 different chemical compounds have been isolated in recent years from the tissues of plants. . . . They are usually not substances used in the regular physiology and metabolism of the plant and they frequently are of bizarre and novel chemical composition. They have thus come to be called secondary compounds. . . . Plants have evolved secondary compounds primarily to discourage herbivory" (*Tropical Nature* [New York: Scribner's, 1984], p. 90). Forsyth and Miyata also refer to the findings of Neal Smith, a researcher working at the Smithsonian Tropical Research Institute in Panama. Smith was raising caterpillars of *Urania*, a beautiful day-flying moth, which he fed with leaves from the *Omphalea* vine, their usual host plant. Smith noted that when he fed the caterpillars leaves from plants that had been damaged by a previous generation of caterpillars, the caterpillars wasted away and died. The *Omphalea* had reacted to the caterpillars' attacks by producing toxic compounds. Smith believes

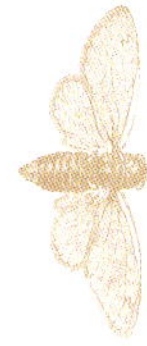

that this phenomenon explains the periodic migrations of the *Urania,* which are certainly not triggered by climatic factors. After a surfeit of caterpillars has aggravated the *Omphalea,* they seem to know, once they have become moths, that they must move to another location.

Tropical species have developed the most extensive chemical defenses of all the plants, no doubt because their life spans are generally longer and tropical insects are more numerous and diversified. The threats that tropical plants face are thus enormous and continue throughout the year.

Since a plant's various parts are not of equal importance, it sometimes localizes its chemical defenses in specific areas. These chemicals may even be transferred to other areas, shifting from the plant's older leaves to its younger shoots, for example. This phenomenon depends not only on the plant's areas of vulnerability, but on the type of insect provocateur that forces the plant to mobilize its defenses.

For a variety of reasons, numerous plants, especially in the tropics, produce new leaves that are not green. Whether violet, pink, or off-white, such leaves are never attacked by insects since it is the color green, conferred by chlorophyll, that signals edibility. This apparently effective defense mechanism relies on a range of colors that have a deterrent effect; leaves with such unusual colors momentarily redirect the attention of herbivores. Given that many defense mechanisms work on more than one level, such leaves may also contain toxic substances, assuring the plant a necessary respite for its survival.

The poisons secreted by plants in response to various threats are often quite specialized. Some plants, such as the tropical vines whose roots yield the compound rotenone, produce powerful insecticides that have little or no effect on herbivorous mammals. According to Forsyth and Miyata, the toxicity of a particular plant species varies from place to place depending on the level of herbivory in that location.

To appreciate one of the most remarkable chemical means of defense, we must revisit the neotropics, where several hundred species of passion vines (Passifloracae) grow. These climbing plants are well defended by the cyanide-related compounds that they produce. However, they also play host to the Heliconidae, a family of butterflies indigenous to the entire neotropical region, which devour the plants' leaves with impunity during the larval stage. The adult butterflies retain the plant's toxins and find themselves protected against any bird that might be of a mind to taste them. The

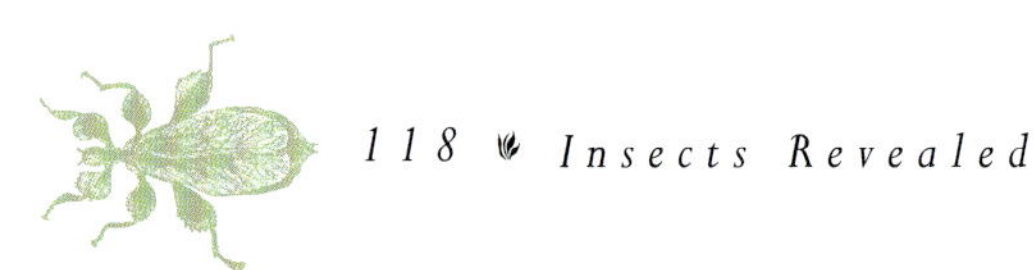

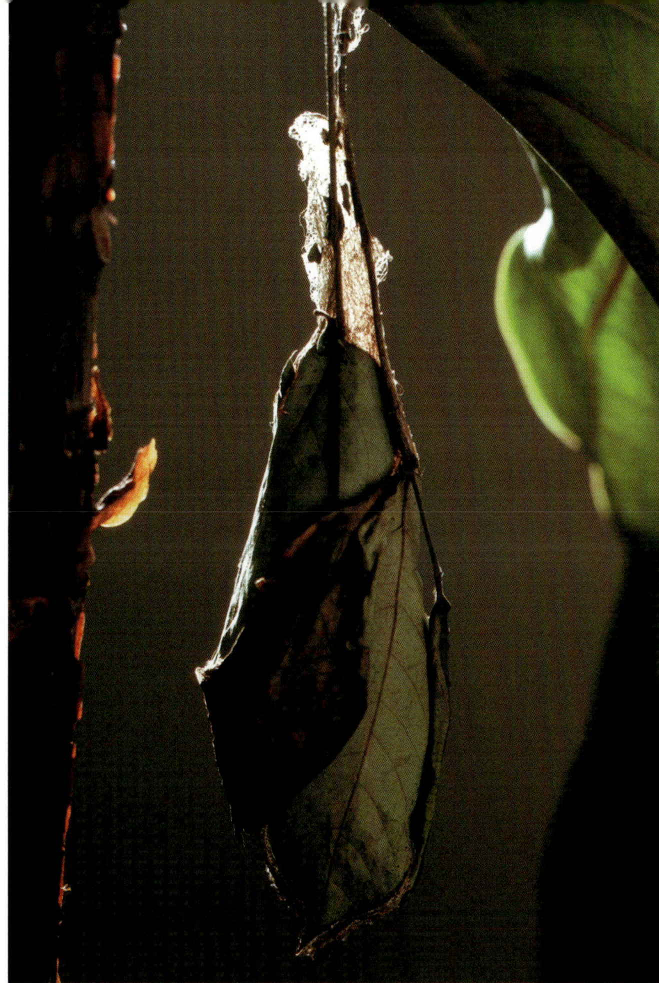

Within the plant world that surrounds them, insect species find either ready-made homes or materials that they can make into shelters.
Above: A cocoon of the atlas moth *(Attacus atlas)* from Southeast Asia. Only moths use cocoons to cover their chrysalises.

Heliconiidae are highly recognizable by their shape, bright coloring, and lackadaisical flying style; the bird will undoubtedly remember the encounter and will have thus received a lesson in warning coloration.[7] In the same way, the monarch butterfly uses the milkweed's chemical defenses for its own purposes; most other foul-tasting or foul-smelling insects, however, produce their toxins themselves.

The Heliconiidae, Danaidae (the monarch butterfly's family), and other butterflies that feed on toxic leaves are protected throughout the caterpillar, chrysalis, and adult stages. Other species that imitate these butterflies enjoy comparable protection—although only in the adult stage—since predators that are forewarned will refrain from attacking and will not take the time to distinguish between real and "false" varieties (see the discussion of mimicry in chapter 12).

Of all the ingenious defenses used in the ongoing war between insects and plants, one of those developed by a neotropical passion vine is particularly impressive. On each of its leaves, this plant produces a tiny bump that resembles the egg of a helicon-

---

7. Such bright colors and patterns are linked with the chemical defenses used by insects, frogs, and other small organisms. Predators quickly learn to decode the signal and to strike these poisonous "untouchables" from their menu.

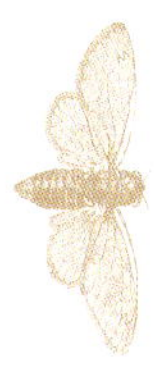

ian butterfly. These structures secrete a sugary liquid that attracts certain ants. The fierce ants defend these nectar sources against all intruders, thereby defending the entire plant.

This first line of defense is not as extraordinary as it might seem; such mutual defense systems are common in the tropics. What is extraordinary is how the plant uses its nectaries to resist heliconian caterpillars. Plants cannot do much once caterpillars are established on them; but to prevent the initial invasion, the passion vine employs one of the cleverest tricks in the plant kingdom. Indeed, its egg-shaped nectaries might lead us to conclude that the plant had discovered the enemy's family secret long ago and subsequently put this knowledge to good use. The secret is this: heliconian caterpillars are prone to cannibalism from birth. Consequently, the mother instinctively (and prudently) lays only one egg per leaf. Upon seeing nectaries in the shape of her own eggs on the leaves, she bypasses the plant and looks for another site.

In addition to food and toxic substances, the plant kingdom also offers insects shelter in every season. Wherever eggs, larvae, chrysalises, or adult insects respond to temperature and humidity extremes by becoming dormant or quiescent, they find either ready-made homes or materials that they can use to make into a shelter.

Plants also serve as models for mimetic insects, which camouflage themselves as leaves, twigs, flowers, bark, thorns, or buds either to escape the watchful eye of predators or to become hidden and fatally effective predators themselves, as in the case of certain flower-petal mantises that lie in ambush on appropriately colored flowers.

In rarer instances, plants may model themselves on insects. For example, *Ophris insectifera,* a neotropical orchid, is pollinated only by certain species of bees; the resemblance of its flowers to female bees is so striking that the male bees are invariably duped. They engage in furious battle with these female impersonators, which coat the bees with pollen that is destined to pollinate the next tempting orchid.

The shared lives of insects and plants largely transcend the all-too-static connotations of the term "coexistence." In fact, the process is one of coevolution of two groups that mutually influence each other to various extents and in various ways. This process is sometimes gentle but more often than not it plays out in a slow and silent war.

insects and their coloration

One important feature of insects' morphology is their color. Color is but one element of form, and in some cases it is secondary to an insect's overall appearance. Thus the stag beetles of the family Lucanidae and the large-sized scarab beetles of the subfamily Dynastinae are more noteworthy for their sculpted mandibles or for their cephalic and thoracic horns than for their coloration, which is typically dark and unattractive. Although these insects are usually nocturnal, their coloration does not serve to camouflage or conceal them during daylight hours as they have other means of disappearing from view.

Coloration usually has an expressive function that is just as important as other aspects of form; but in all cases of iridescent coloration, which produces a metallic shimmer, the dazzlingly brilliant colors surpass other features in overall importance.

Regardless of the combinations of colors, shapes, sizes, and textures exhibited by insects, and no matter how delightful we may find these characteristics, they are not intended for a human audience. Instead, these features make up a system of signs allowing members of a species to recognize each other and, in some cases, serving as a warning to their predators not to touch them.

In general, the coloration of adult insects is determined when they hatch or shortly thereafter. When a butterfly emerges from its chrysalis, its soft and shriveled wings already display their definitive coloring, but when a beetle sheds its pupal skin, the chitin present in its outer covering (or integument) does not yet possess the pigmen-

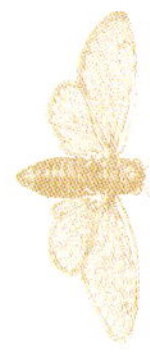

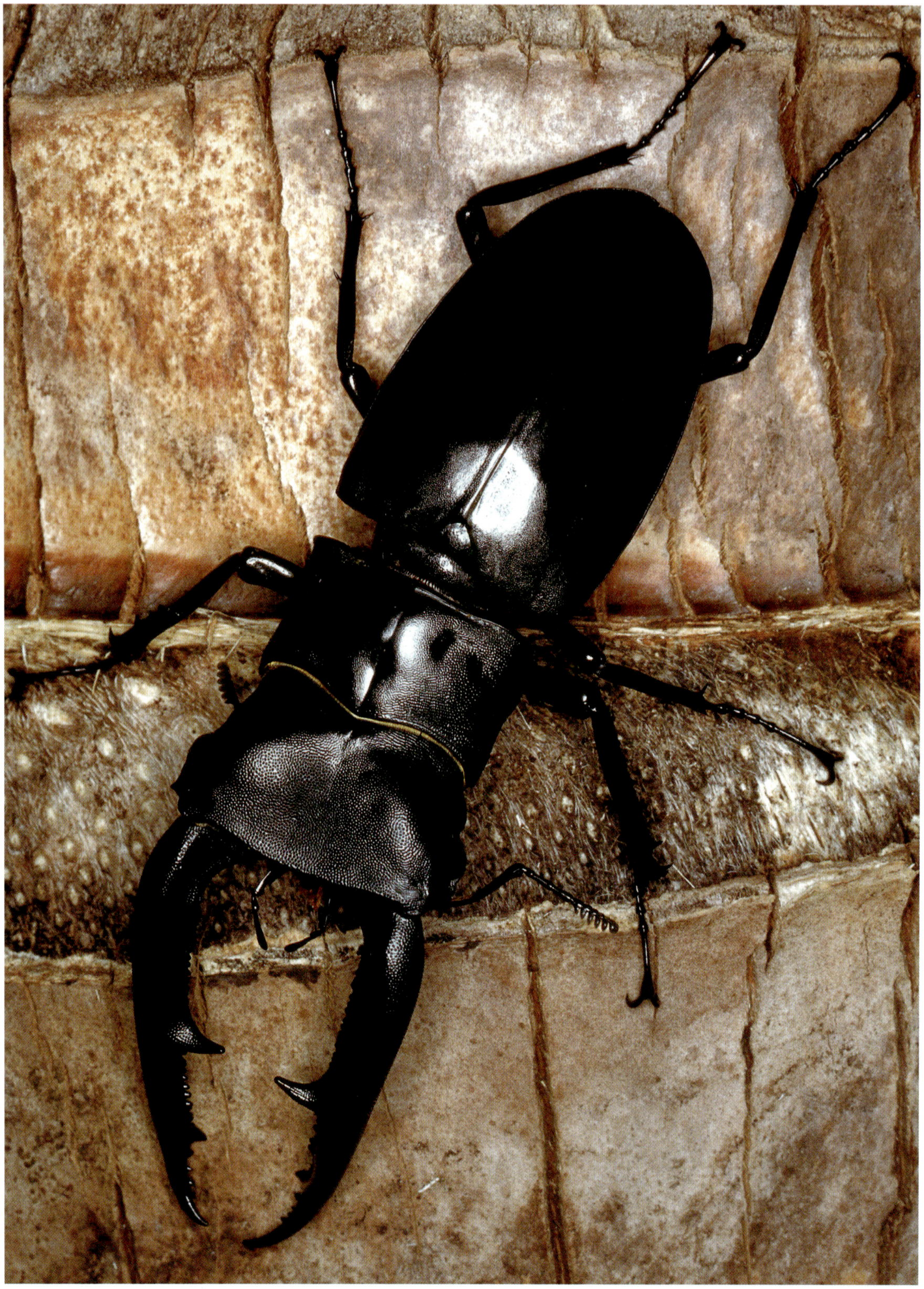

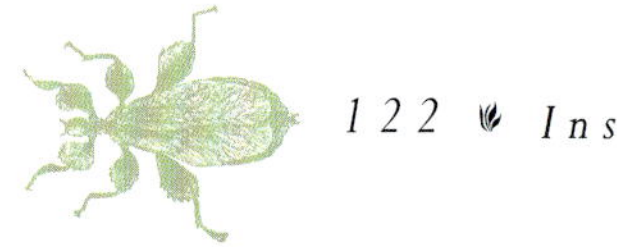

Coloration is only one aspect of insects' form and is often of secondary importance in their overall appearance. Many stag beetles, such as this *Hexarthrius mandibularis* from Sumatra, are more noteworthy for their sculpted mandibles than for their coloration.

tation that will come to characterize it later on. For several hours the off-white and semi-transparent beetle looks like a ghostly member of its own species.

The larvae of insects that live underground or in wood that they consume have pigmentation only in the hard body parts such as the mandibles and tarsal claws.

Numerous species of caterpillars go through quite different stages of coloration from one molt (or ecdysis) to the next; some with green coloring, which serves to camouflage them against a background of foliage, will go on to adopt brownish hues when seeking a suitable support for pupation.

The phenomenon of color variation among migratory locusts is related to population increases. These insects exhibit camouflage coloration in their solitary and quiet phase, and then go on to display vivid colors, social gregariousness, and a strong urge to migrate as a result of the stimulus of living in abnormally closely packed groups.

Each type of insect may be characterized by predetermined color motifs; nonetheless, there are color variations specific to generations of butterflies born in the spring or the fall or, in tropical regions, in the dry or wet season.

An even more exceptional phenomenon occurs among insects (and possibly in other parts of the animal kingdom): certain species of the butterfly family Papilionidae exhibit female polymorphism, in which females differ from each other as well as from males. In the case of *Papilio dardanus,* found throughout eastern Africa, females of the species mimic a toxic danaid of their local area and bear no resemblance to the males of their own species; in Madagascar, males and females of this species are virtually identical. This is akin to female lions being jaguar-like in one region, tiger-like in another region, and "normal" in yet another region.

Such intraspecies variation is more frequent in the tropics than in the temperate zones. The polymorphism of the Heliconidae proves bewildering even to expert taxonomists since virtually every species has given rise to numerous regional forms that are so different from one another—and from the typical representative of the species—that an observer could easily think that they constituted different species altogether.

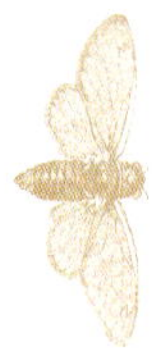

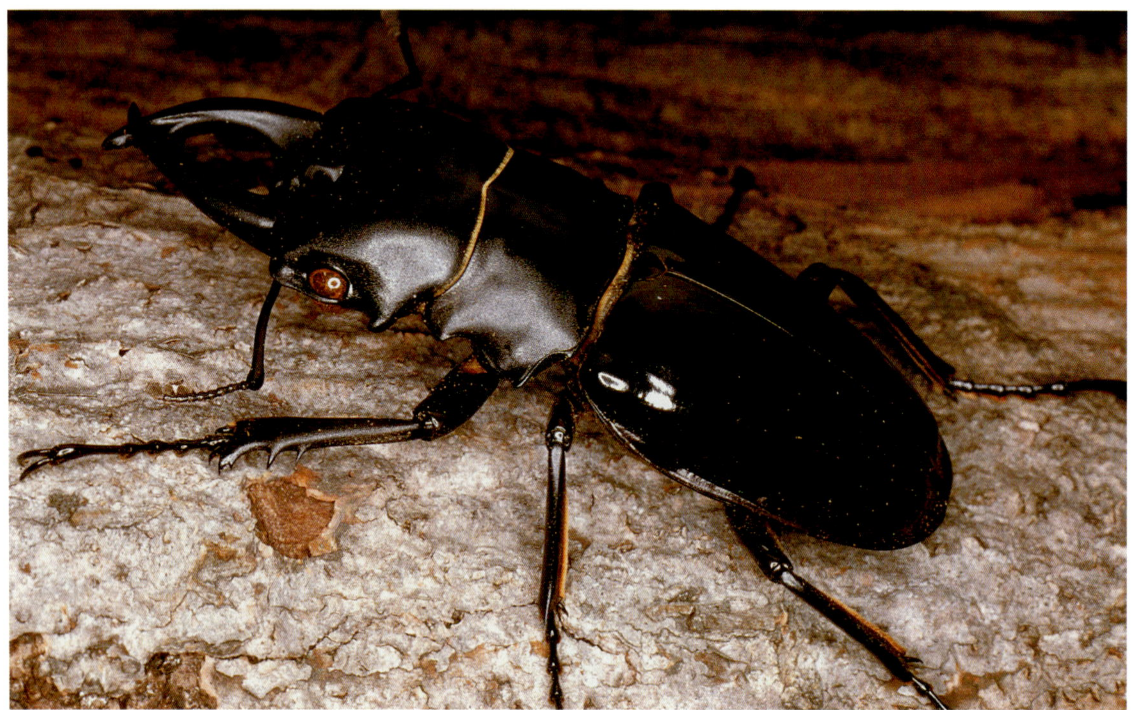

I observed a different kind of polymorphism on the western slopes of the Andes Mountains in Ecuador: there, individuals of *Copaxa multifenestrata,* a saturniid moth, appear with different colorations at the same time and in the same place. Of the large number I observed, I collected five specimens, ranging in color from lemon yellow to brick red to chocolate brown. Each variety appears to be well defined and well established, and there are no intermediate forms.

Sexual dimorphism is often expressed as a difference in male and female coloration, together with a difference in size and antennal or mandibular structure, etc.

Insects' coloration may be caused by pigments in the chitinous integument itself or by the superposition of the integument's thin layers; light is refracted across these layers, producing an iridescent metallic shimmer. These structural colors often blend in with a background of pigmentary colors that temper the overall brilliance of the coloring.

Although most insects display their coloration via the pigment present in their chitinous integuments, others (such as butterflies and several types of beetles) are decorated with scales, which, when elongated, become hair-like strands that produce all conceivable color schemes. Beneath this fragile finery, so perfectly designed at birth, lies a uniformly colored integument—brownish-yellow in butterflies and black in many beetles—that the slightest scratch will lay bare.

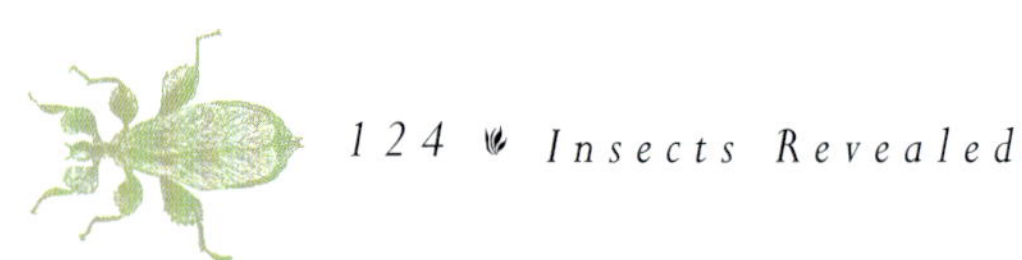

As we have already seen, the Lepidoptera are defined by the scales that cover their membranous wings; without such scales, butterflies' wings would be as transparent as cellophane. Several butterflies from tropical America serve to illustrate this point. Certain Satyridae, numerous Ctenuchidae, and some Ithomiidae have wings with very few scales; this renders them almost invisible in the dimly lit rainforest regions. In the world's temperate climates, few butterflies have transparent wings, with the exception of hummingbird clearwing hawk moths, which are diurnal. In fact, transparent or semi-transparent wings are found more frequently in diurnal than in nocturnal Lepidoptera; among the latter, however, several species of Saturniidae have wings with eye- or crescent-shaped transparent spots, the effects of which are simply extraordinary.

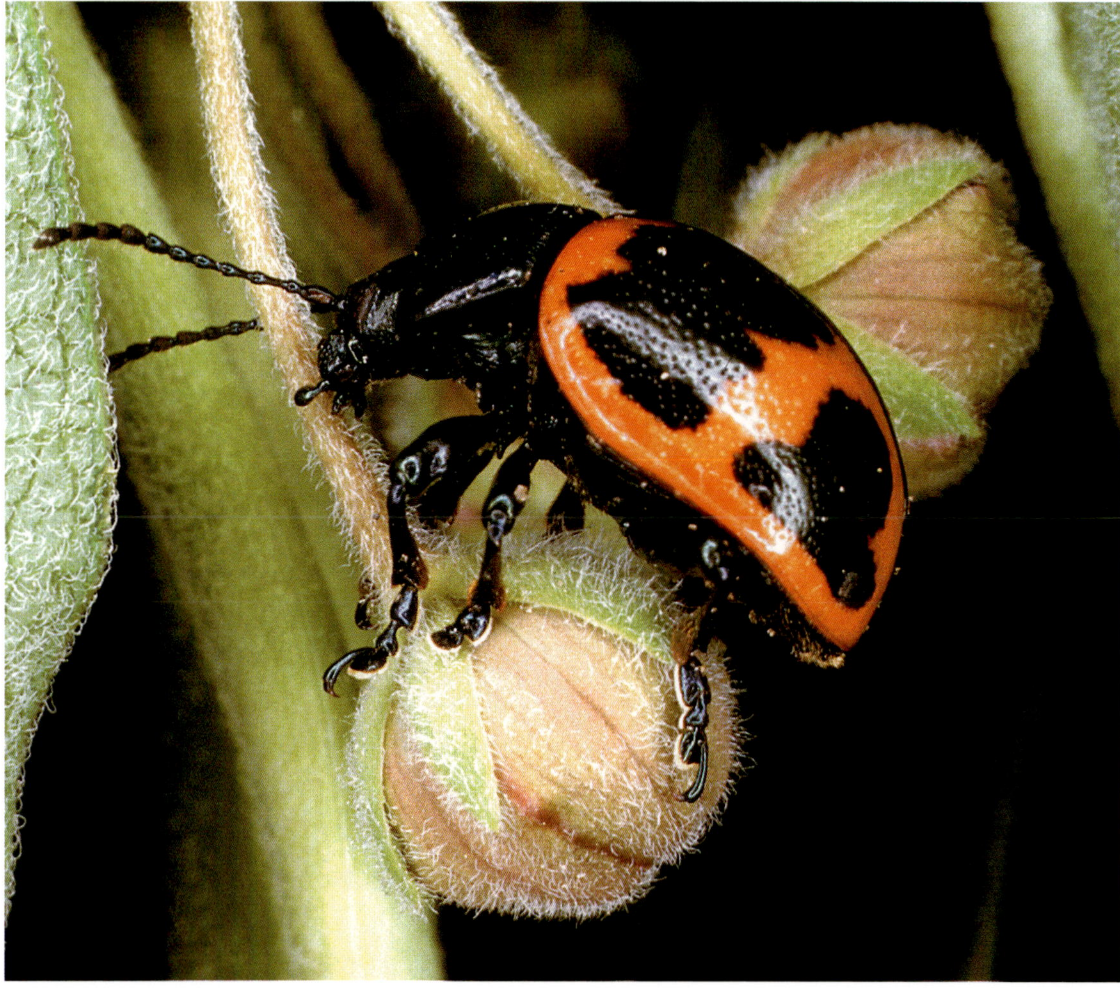

The coloration displayed by the vast majority of insects is due to pigmentation within the chitinous integument; the surface texture can either enhance or temper the color.
Above: An example of chitinous pigmentation. A milkweed leaf beetle (*Labidomera clivicollis*, family Chrysomelidae), indigenous to North America, on a milkweed plant.
Opposite page: A stag beetle from Southeast Asia (*Odontolabis celebensis*). The same black pigment can produce either a dull or a shiny finish.

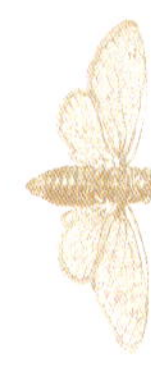

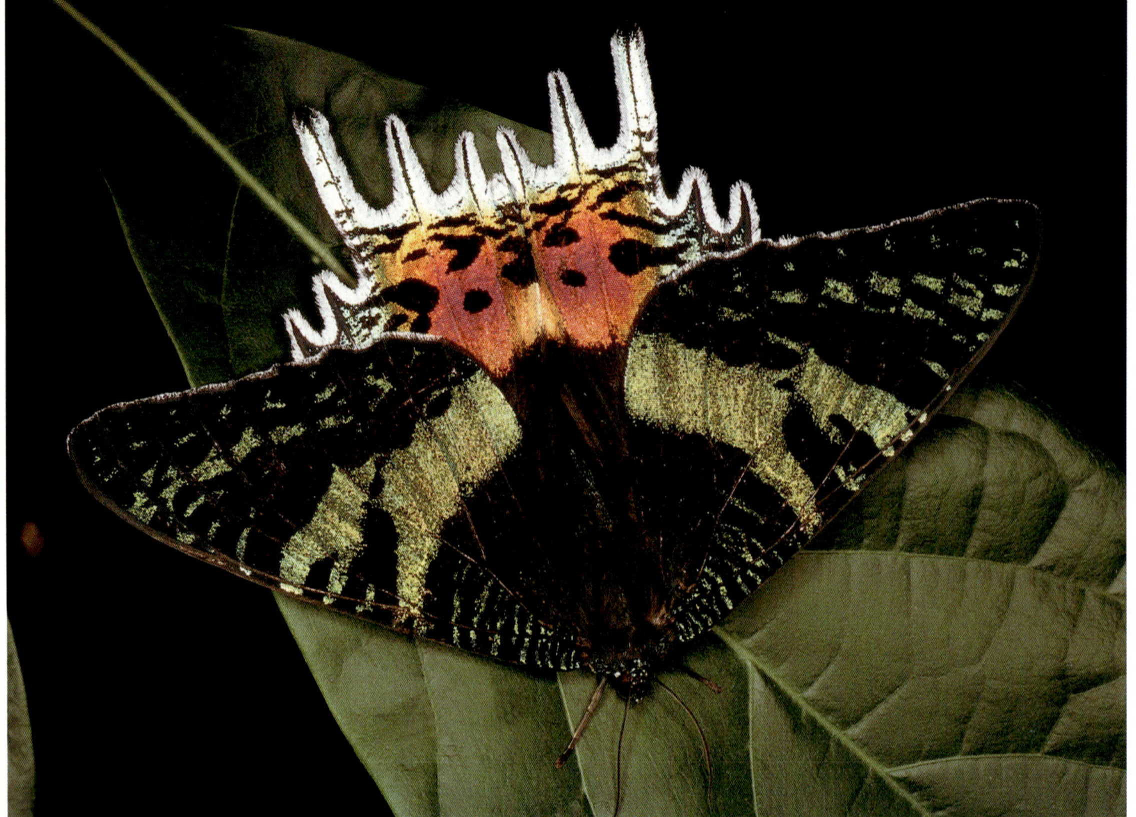

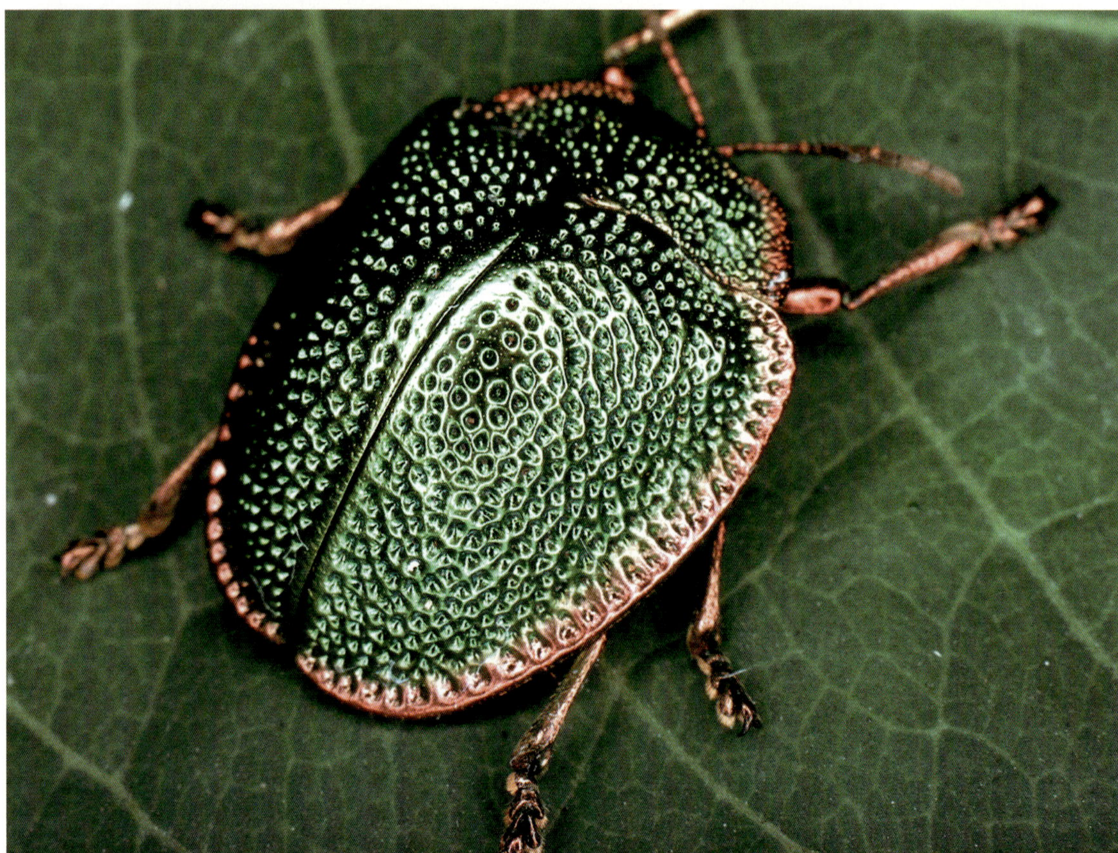

Top: Among specimens such as the sunset moth *(Chrysiridia riphearia)* from Madagascar (a diurnal uranid species), coloration is determined by the structure of the scales covering the membranous wings.

Bottom: Among beetles and other insects with neither scales nor hairs, iridescence is an integral feature of the integument. Shown here: *Desmonota variolosa,* a chrysomelid beetle (subfamily Cassidinae) from Amazonia.

Opposite page: Posterior wing from the sunset moth. The iridescent scales contrast with the black pigmentary scales.

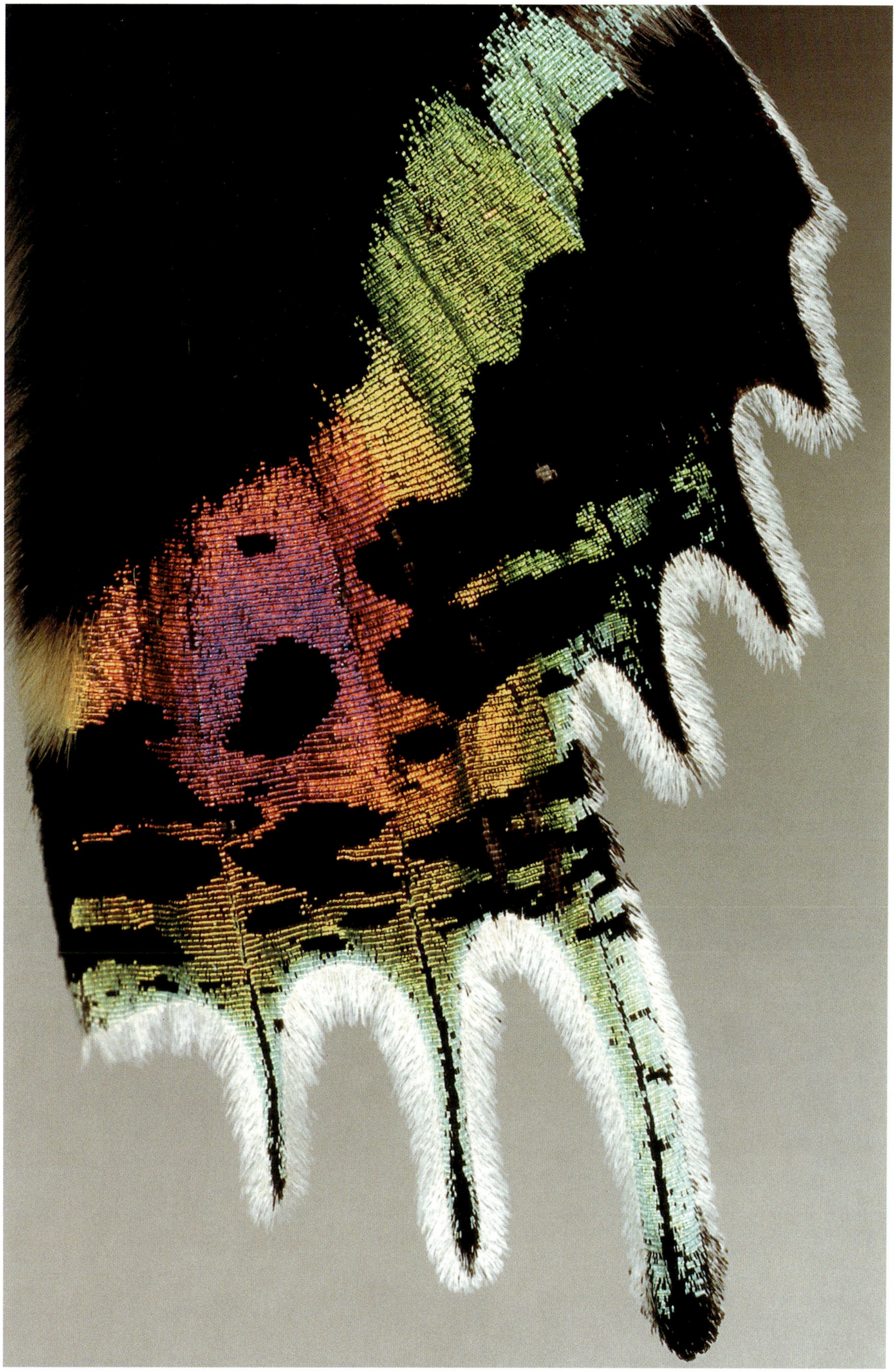

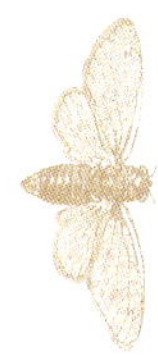

Iridescence: Breath-taking colors of the tropics
This metallic coloration results from the superposition of fine transparent layers that break light into dazzling colors that no pigment could ever produce.

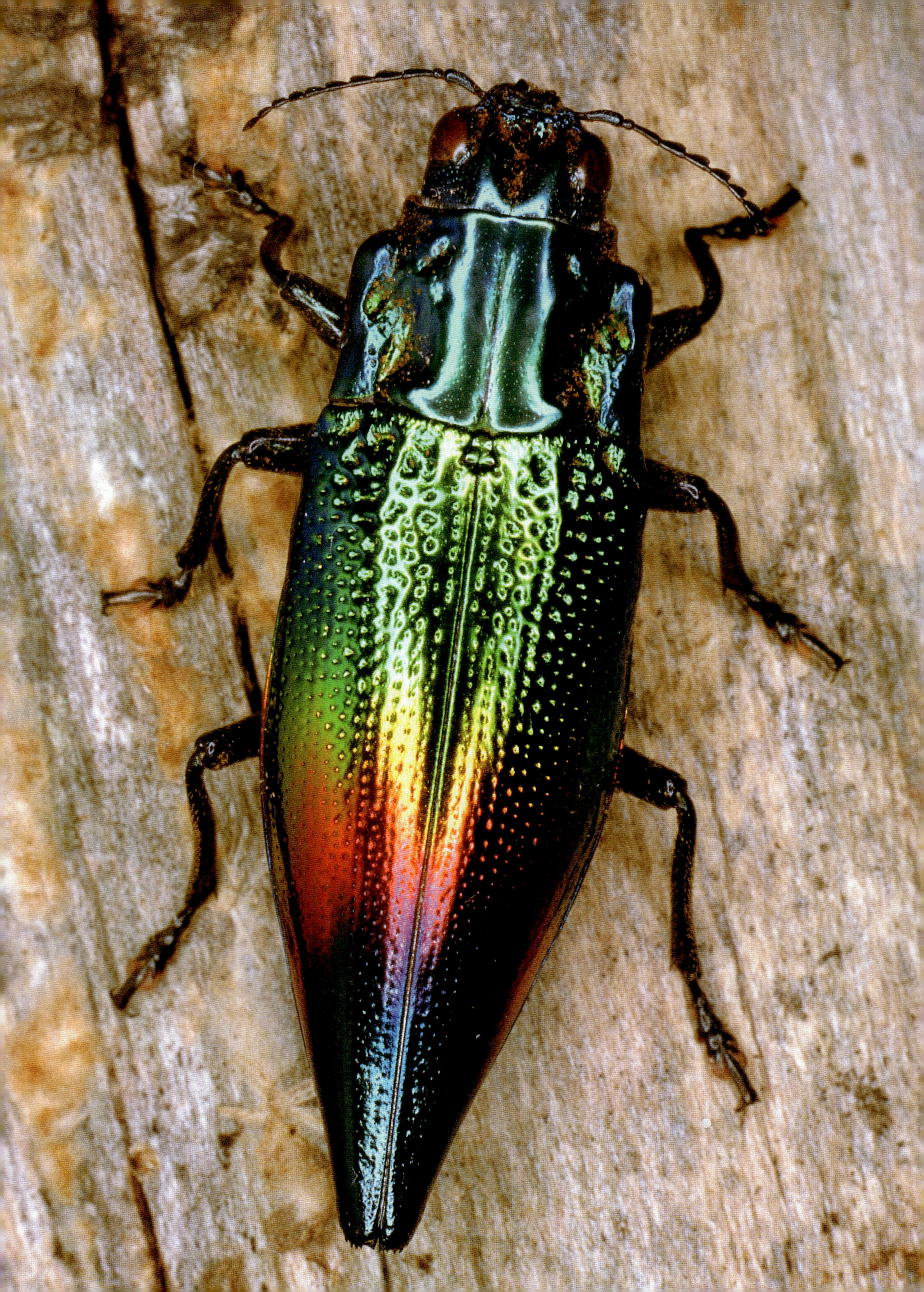

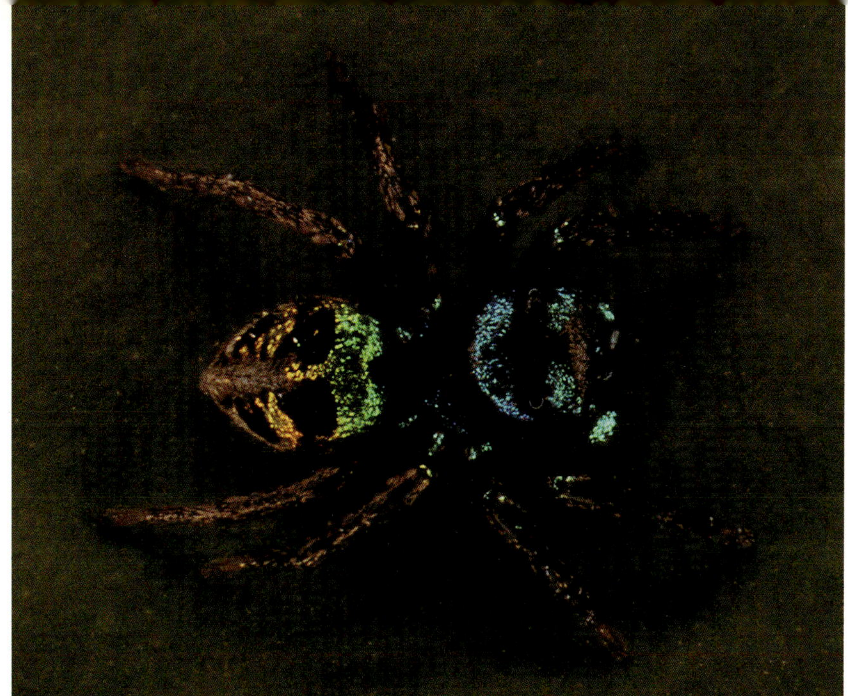

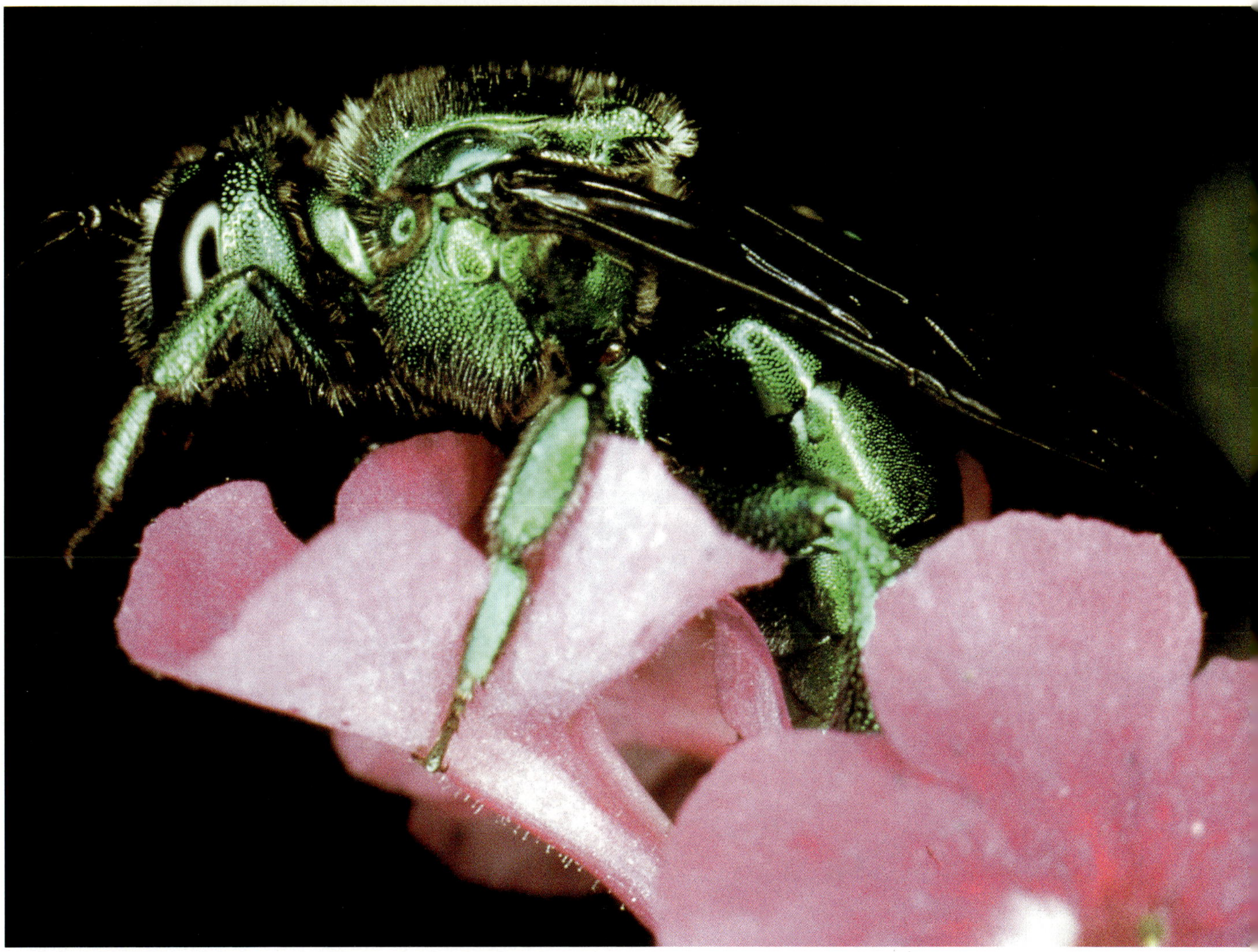

Top: Insects are not the only creatures to sport iridescent colors. Other organisms, such as this small jumping spider from the Dominican Republic, display iridescence as well. However, nowhere else is this phenomenon as widespread and magnificent as it is among the Lepidoptera (butterflies and moths), Coleoptera (beetles), and Hymenoptera (bees and wasps).

Bottom: Euglossine bee from the Amazonian region of Peru.

Opposite page: This brilliantly colored tropical beetle in the genus *Cyphogastra* is a buprestid from Papua New Guinea.

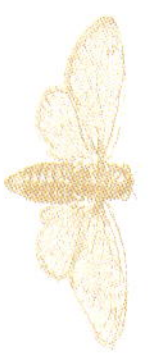

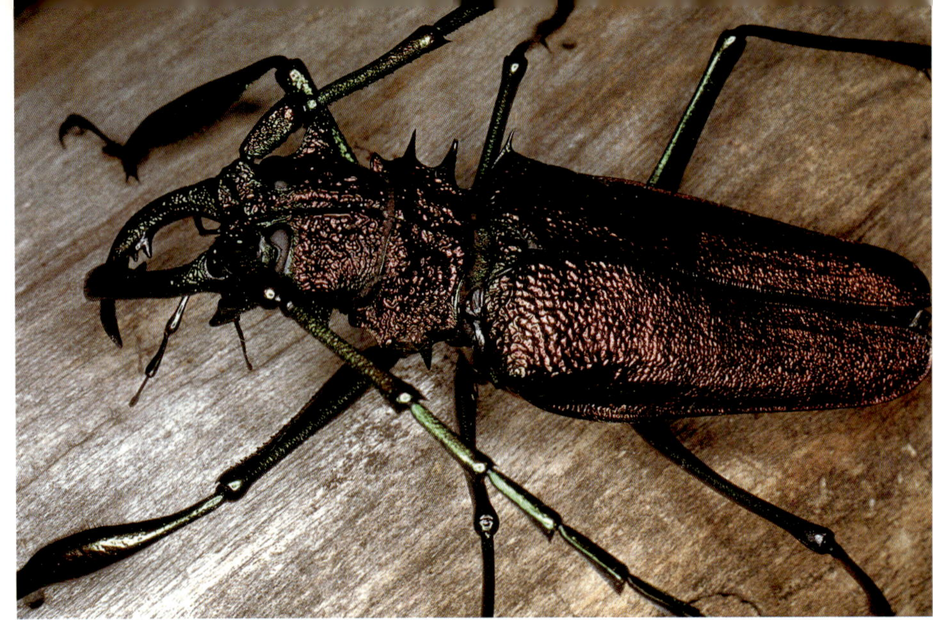

Nature abounds in signals relating to color, light, smell, and so on, which, to reiterate, are not intended for a human audience. Human beings are the only species preoccupied with the communicative value of signals that do not directly affect them. We are curious about such signs and, at times, our naïveté influences our interpretations of them.

A lioness on the plains of Kenya pays absolutely no attention to the mating calls of a nearby cricket, just as a butterfly remains impervious to the pheromones emitted by males or females of a different species.

For all species other than our own, such signals are received almost exclusively by the members of the same species, although at times members of other species may also be party to the signals transmitted. These signals constitute intraspecific communication when the coloration of a species allows its members to recognize each other, and interspecific communication when the coloration also sends a warning to predators.

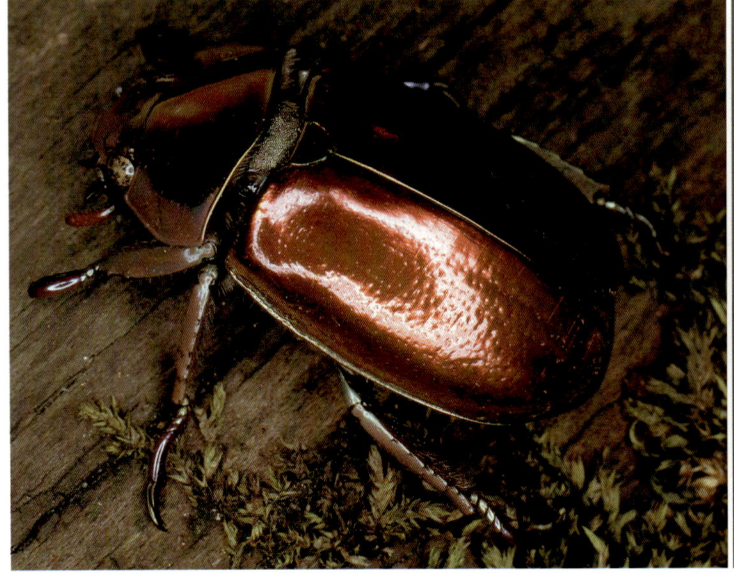

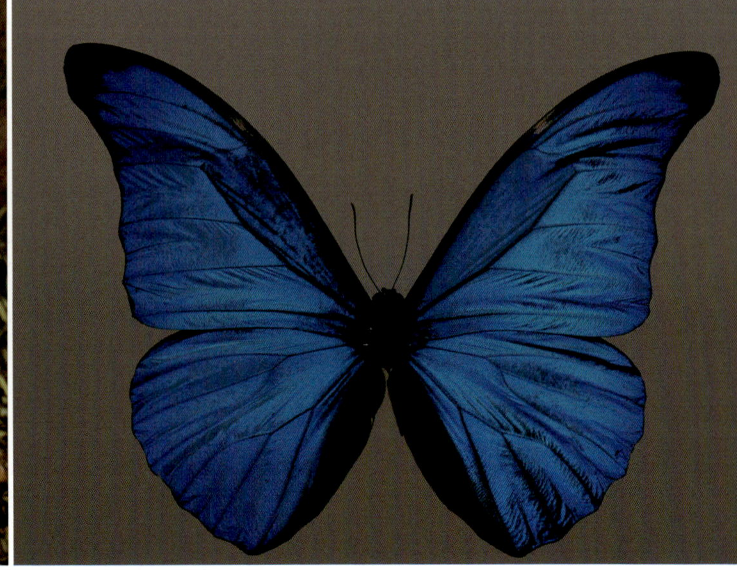

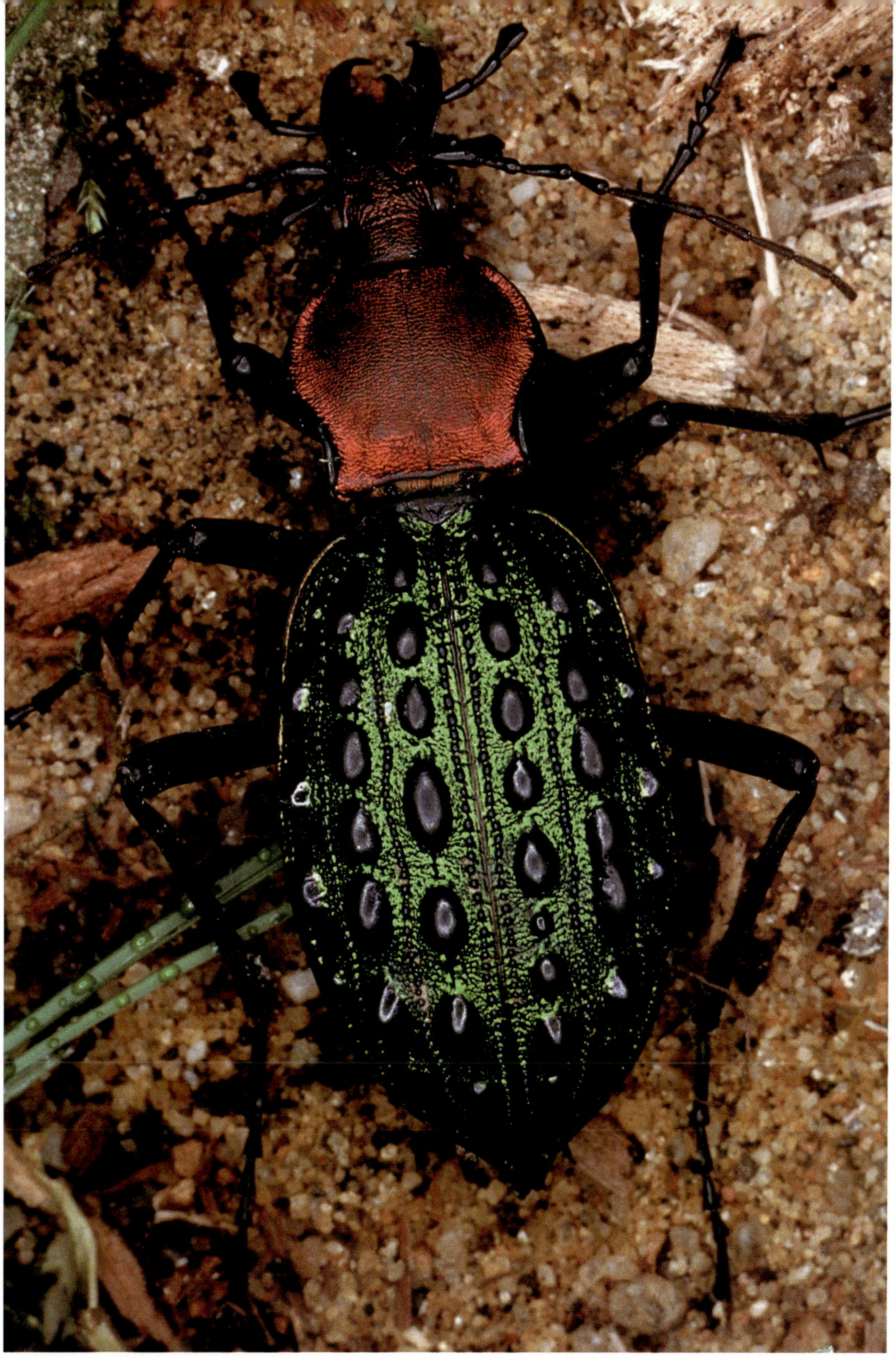

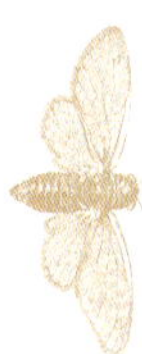

Above: The carabid beetles around the world outdo each other in the brilliance of their color. *Coptolabrus pustulifer,* from China, is a fine example.

Opposite page, top: Most cerambycid beetles in the subfamily Prioninae look quite similar to each other. The genus *Psalidognathus* contains several remarkable exceptions, such as *Psalidognathus friendi,* which has exchanged standard black and brown for this brilliant finery.

Opposite page, bottom left: The scarab beetles in the genus *Plusiotis* (subfamily Rutelinae) feature some of the most glorious examples of iridescent coloration. The golden beetle *(Plusiotis aurigans)* comes from the Talamanca mountain range in Costa Rica.

Opposite page, bottom right: Although it makes little sense to hold a beauty contest for insects, if there were one, many would give their vote to the morpho butterflies of tropical America. *Morpho rhetenor* is widespread in northern South America.

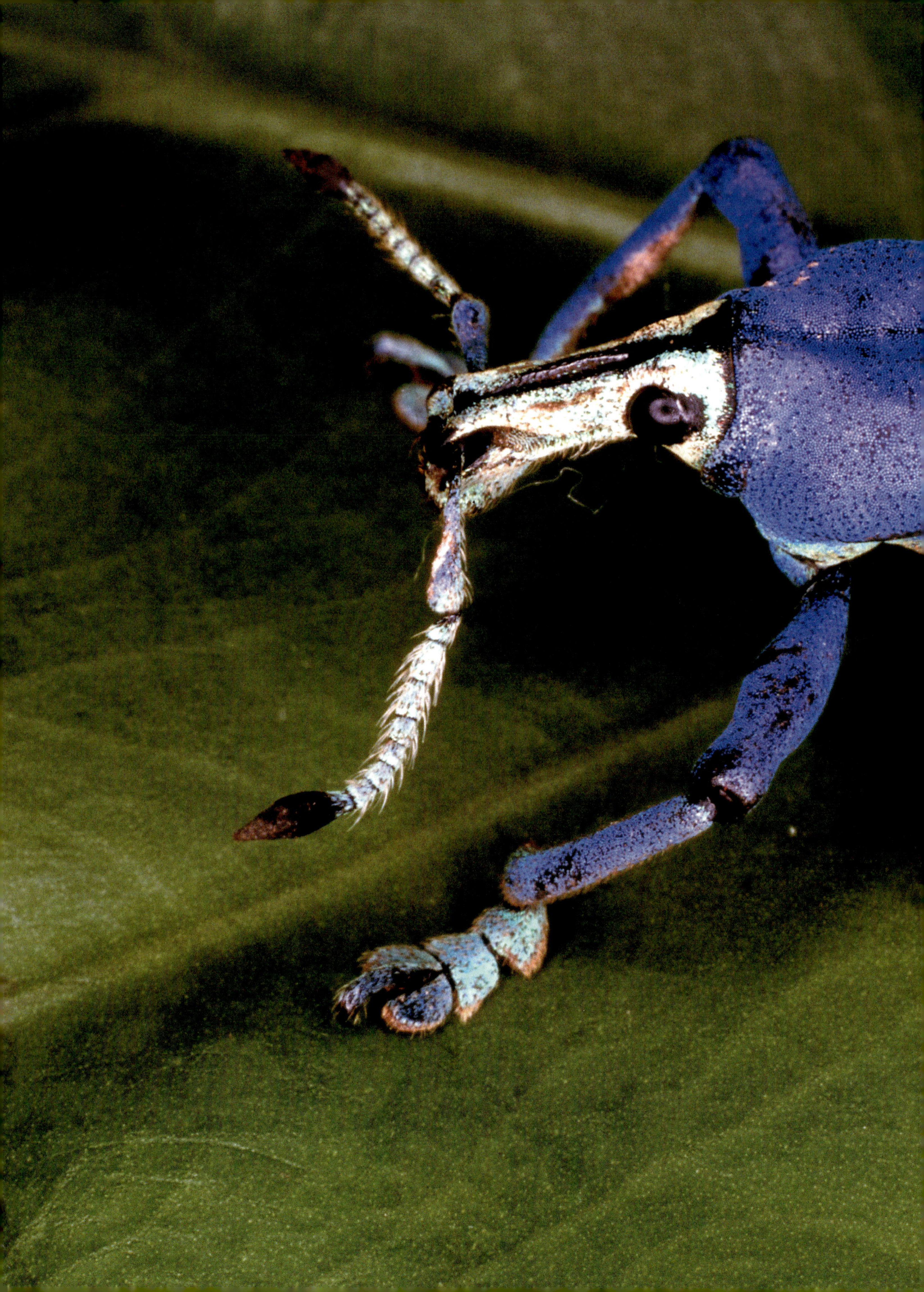

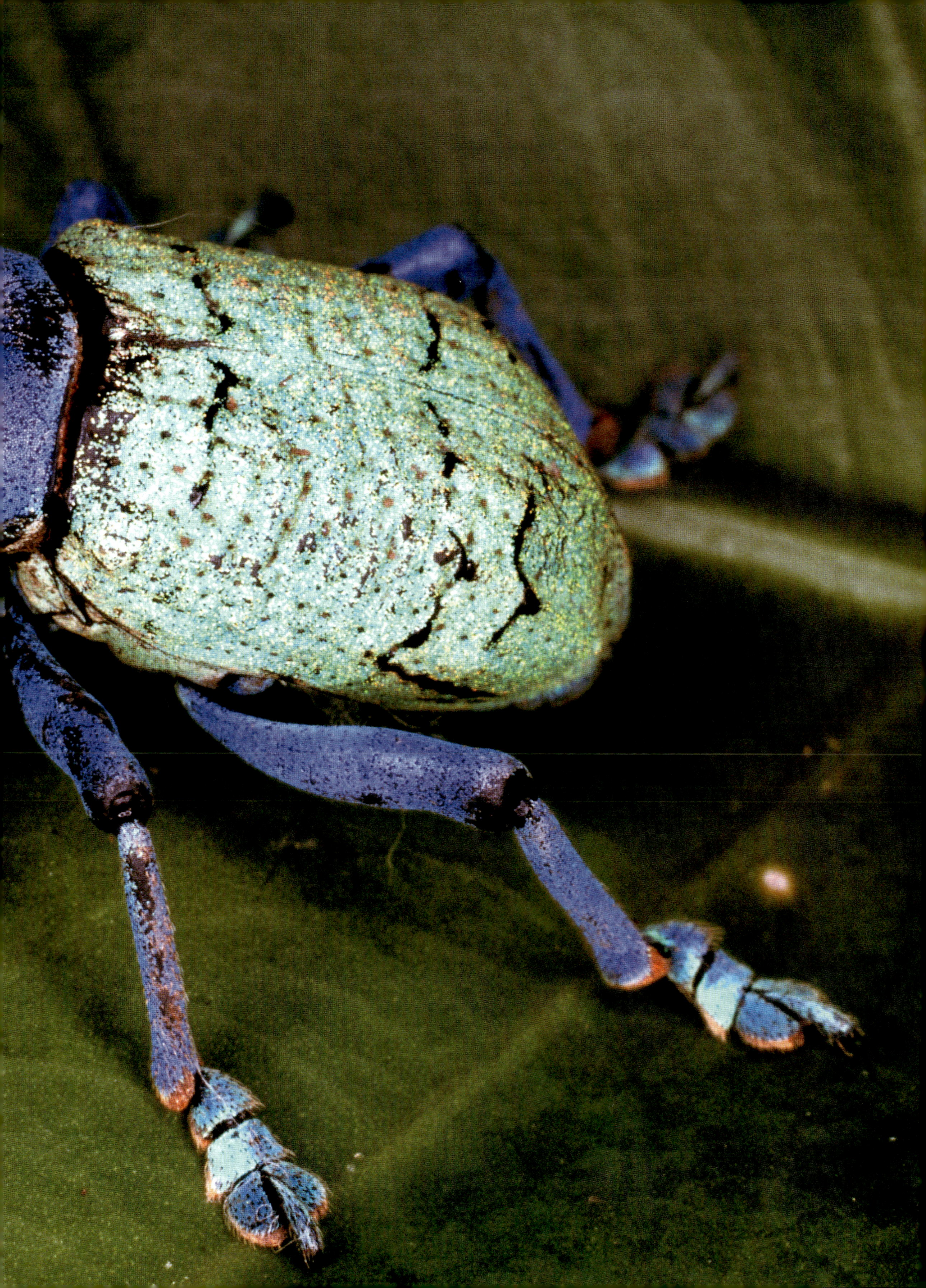

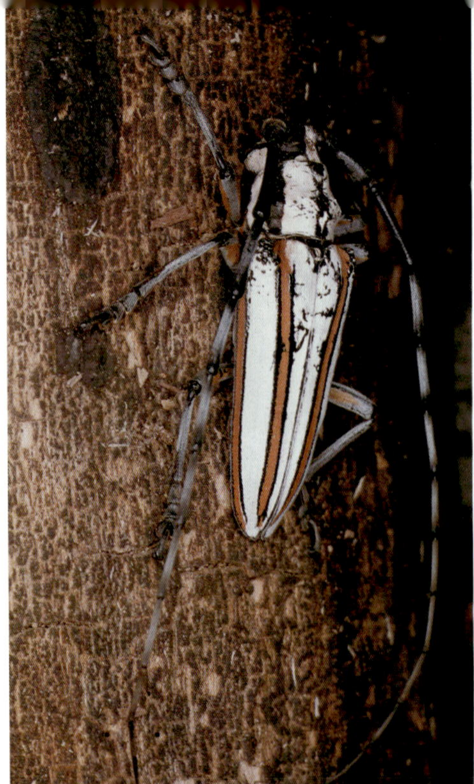

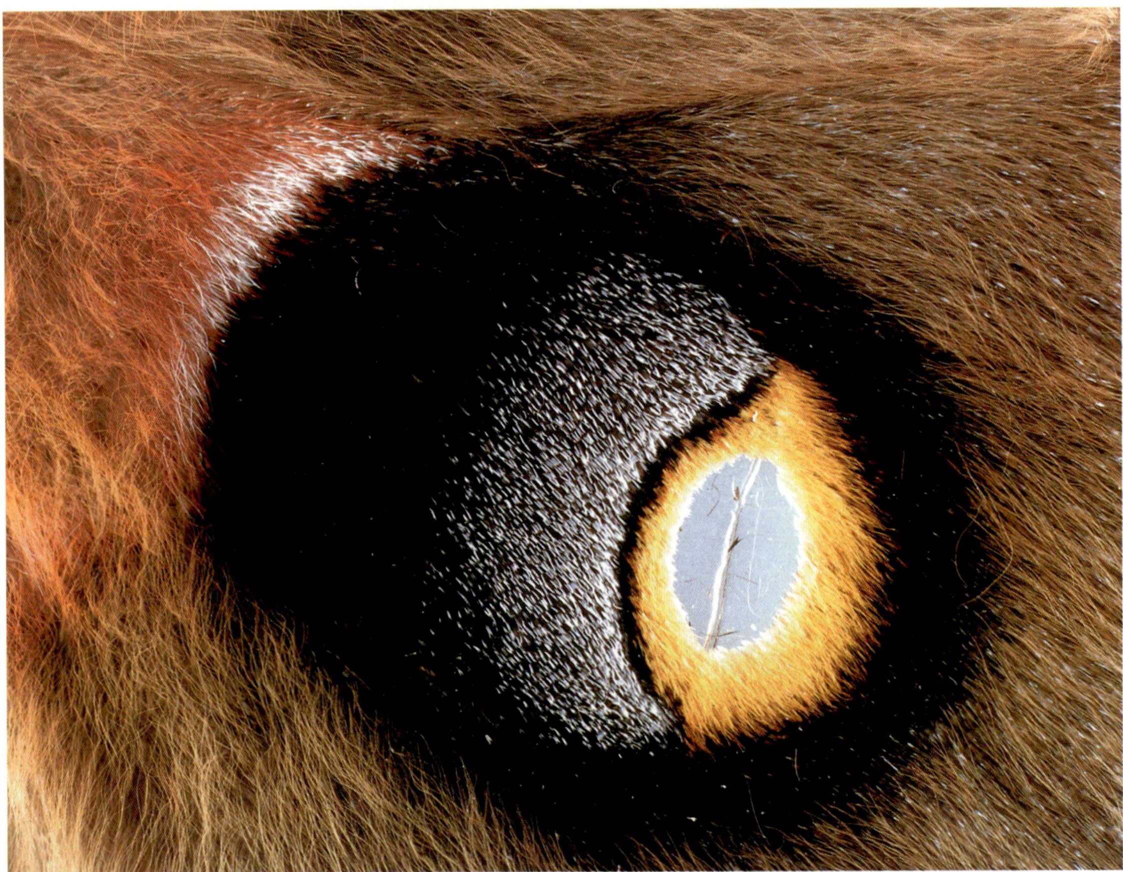

Most insects display their coloration in the chitin of their exoskeleton. The Lepidoptera and some others, however, are covered with elongated, hair-like scales that form a wide variety of patterns. This covering is fragile; the slightest scratch exposes the underlying integument, which is most often brown or black.

Top: The effects of scratches on this beautiful Mexican cerambycid beetle (genus *Deliathis*) are clearly visible; the base color appears as black against the white of the elytra.

Bottom: Hairs covering the posterior wing of the American polyphemus moth *(Antherea polyphemus).* The transparent membrane is visible where the ocellus forms a small window.

Preceding pages: *Eupholus schoenherri semicoeruleus* from Papua New Guinea, a beetle in the family Curculionidae, is also covered with scales that produce its color.

12
insects' defense strategies

CAMOUFLAGE AND MIMICRY

The scenarios described in this chapter do not bear witness to the intelligence of individual organisms; instead, they are signs of reciprocal relationships between partners caught up in escalating aggression and defensive maneuvers. Nevertheless, it may be helpful to imagine that these behavioral aspects are the product of some form of intelligence; indeed, insects' extraordinary defense strategies appear to be endowed with an admirable sense of purpose.

Consider this phenomenon: many insects resemble various elements of the plant kingdom, or even clumps of earth or bird droppings. We understand how natural selection may serve to reinforce such mimetic or imitative characteristics with what seems an astounding intentionality once the process is underway. But what triggered this process in the first place? How did an Amazonian longhorn beetle ever come to camouflage itself as bird droppings? How did the forerunners of the Southeast Asian leaf insects ever begin to resemble leaves? More than a century after Darwin, these questions point to this phenomenon's abiding mystery.

Of all possible defense strategies, the most economic and effective is simply to disappear from sight. This strategy, by which an insect resembles through form or color some object that holds no interest for a predator, is known as cryptic coloration (from the Greek *kruptos,* concealed).

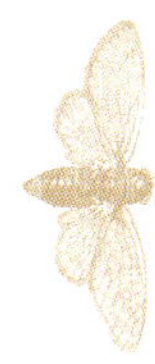

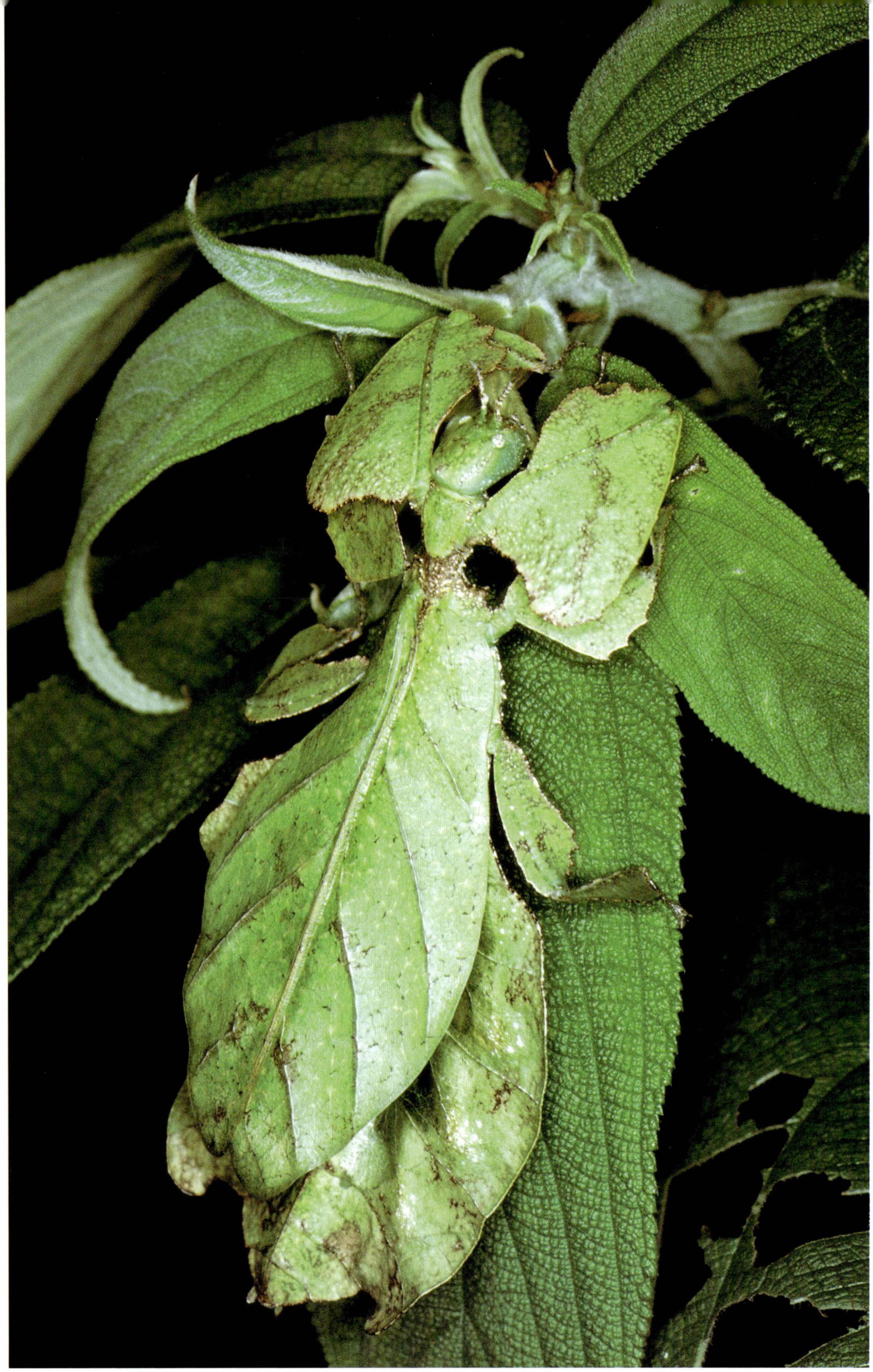

Nature seems inventive not only in creating primary forms, but also in counterfeiting them. As in the art world, there is both authenticity and forgery. Nevertheless, these "forgeries" are as extraordinary as the originals. This leaf insect wins the contest among those insects that model themselves on plants in order to disappear into the environment; their existence is as peaceful as it is deceptive.

This leaf insect *(Phyllium siccifolium)* is a member of the order Phasmida (from the Greek *phasma,* ghost). It is found in Southeast Asia together with fifty related species. This order also includes the walking-sticks, which are widespread throughout the world.

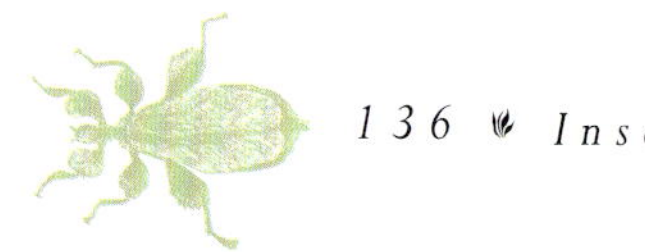

Unlike certain fish that shift their coloration from pale to dark according to their visual perception of variable aquatic environments, insects relying on cryptic coloration must find surfaces that match their camouflage colors, which are usually invariable. Under normal conditions, these insects camouflage themselves instinctively.

Because nature's forms and colors are indivisibly connected, the phenomena of matching forms and matching colors always complement each other. Cryptic coloration thus implies form as much as it implies color.

Cryptic coloration constitutes a passive and precautionary form of defense. It is the greatest of luxuries if an insect (or any other prey) is able to avoid detection by its predators since it is always safer to avoid unpleasant encounters than to deal with their aftermath.

Virtually all moths utilize some form of camouflage. Certain large neotropical noctuids (genus *Thysania*) flatten themselves in a more or less oblique position on a tree trunk so that the markings of their wings match the striated pattern of the bark. This strategy serves to confuse and thwart birds that might detect the moths if they positioned themselves horizontally. Similarly, some tiny moths sometimes bring their posterior wings in front of their anterior wings, thereby changing their normally triangular resting postures. And some species contort themselves into acrobatic positions, remaining motionless all day long, as if they considered themselves—or wanted others to consider them—to be twigs or shriveled brown leaves.

In all cases of form and color matching, insects ingeniously adopt patterns of behavior designed to make their camouflage as realistic as possible. Here too, a variety of shrewd alterations and clever tricks helps sow confusion in the minds of animal predators and human observers alike.

Among the hundred or so species of neotropical saturniid moths in the genus *Automeris,* several exaggerate their similarity to dead leaves to such an extent that they move like a fallen leaf hanging by a tiny thread, buffeted by indiscernible air currents. To achieve this effect, the moth perches on a tree trunk and brings its front legs together to form a single contact point with the underlying surface. Using this contact point as a pivot, it moves from side to side using its middle legs, which are hidden under its wings, to emulate the motion of the neighboring leaves.

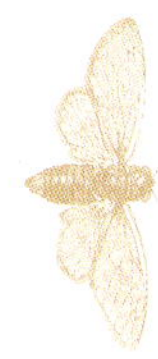

Once it has blended in with its surroundings, the moth remains still. If some especially curious bird then decides to verify the authenticity of this "leaf," the moth has a second defense strategy that it can use before it flies off: it brings its anterior wings forward to reveal its posterior wings, which are decorated with large eye-like markings, described by several authors as "frightening." In anthropomorphic terms, we might say that these markings should scare off predators or, at the very least, startle them. But are these "eyes" as truly frightening as some say? Do birds actually see what we see? Predators learn through experience and are not always taken in by mere appearances. This applies especially to the superb eye-like markings, or ocelli, found on the posterior wings of the large *Caligo* butterflies indigenous to tropical America, also known as the "owl butterflies." Photographs of these creatures are often inverted, with the butterflies' wings spread to show off these distinctive markings; no doubt this owlish disguise should frighten any bird that happens by.

In reality, however, the scenario is less dramatic. Predators never end up confronting this mask-like feature since the *Caligo*s always land with their wings folded over each other; birds thus only see a single one of the butterflies' wings at a time. Nevertheless, when viewed in the half-light of the rain forest, these ocelli must have a significant deterrent effect, since the butterfly seems to put its faith in them: like all organisms that use cryptic coloration to emulate foreign objects, it remains motionless to the last minute, sometimes until the moment of death itself. Every strategy has its own logic as well as its limits.

Among the more prudent kinds of cryptically colored insects are the green katydids of the family Tettigoniidae, which imitate either blades of grass if they make their homes in fields, or leaves if they live in or among trees, as in the case of tropical species. Stick-like phasmids and leaf insects are also experts at mimicry; indeed, giant phasmids may resemble branches the size of one's thumb. All these insects are active by night and their camouflage is only of use during the daytime when they sleep and remain still. Cryptic coloration is ineffective at night since nocturnal predators rely primarily on their senses of smell and hearing to locate their prey.

In the Amazonian regions of Ecuador and Peru, I often saw unusual patches of moss growing close to the base of tree trunks. Taking a closer look, I discovered an amazing case of collective camouflage: caterpillars with nettle-like hairs were packed together so tightly that I could not tell them apart. This strategy is much more complex than

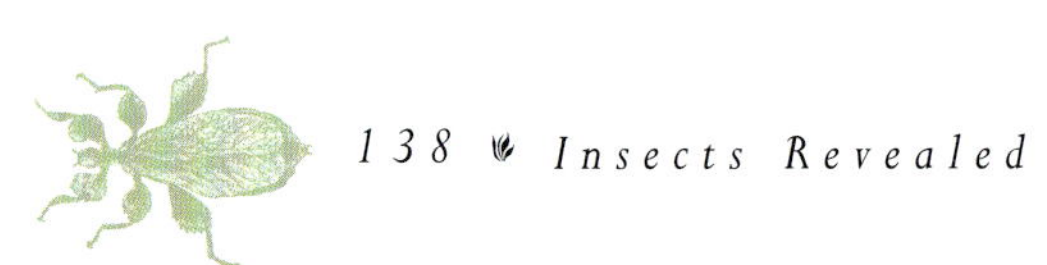

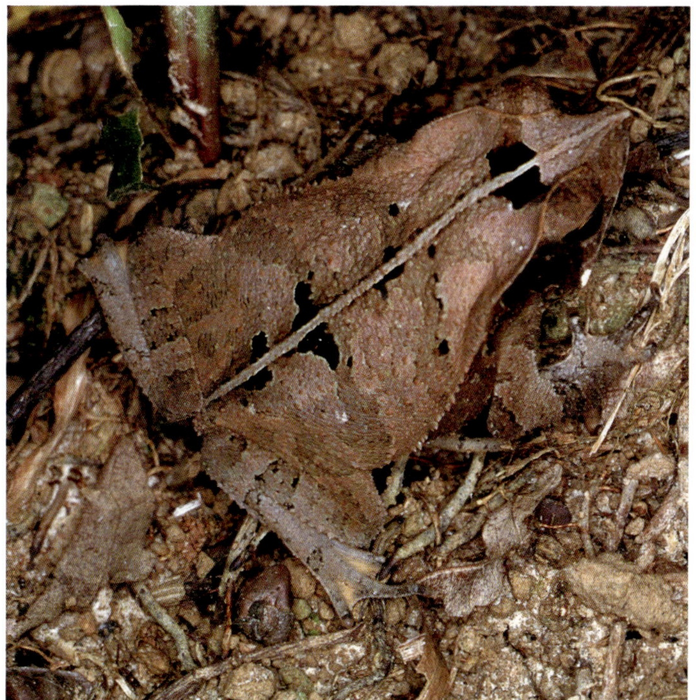

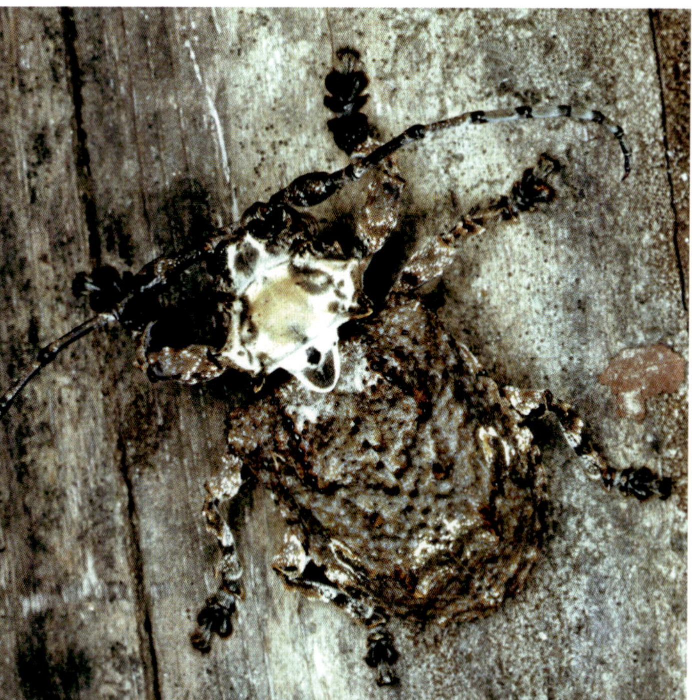

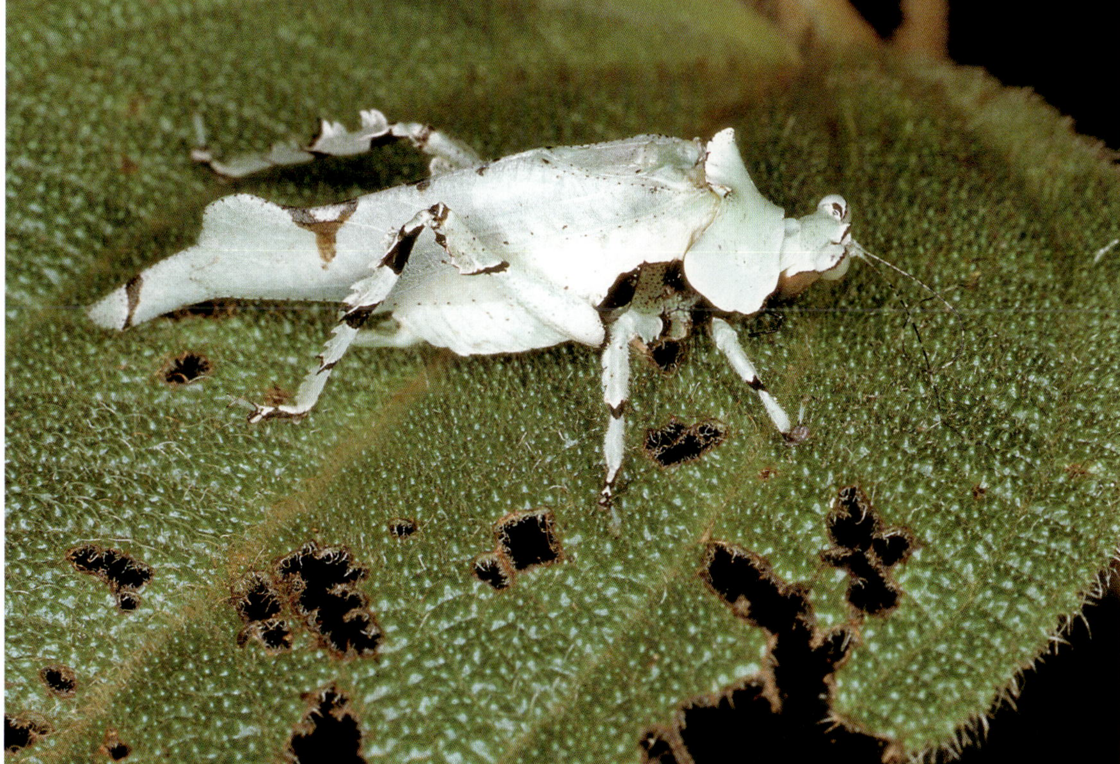

Cryptic coloration: a passive and preventive form of defense
To escape the attention of predators is the greatest luxury for any prey species. Here are three examples of prey whose chances of survival depend on passive defenses. The dead-leaf frog is almost invisible in its environment. *Onychocerus crassus,* a cerambycid beetle from the Amazonian region of Peru, disguises itself as bird droppings, as does this grasshopper from the Amazonian region of Ecuador.

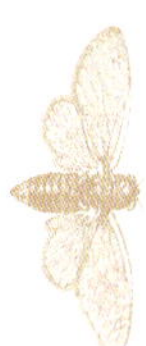

individual camouflage behavior since it requires each member of the group to act in solidarity with the others. At nightfall, while most birds are asleep, these caterpillars resume their grazing high up in the treetops. Several species of caterpillars use this strategy of collective dissimulation to imitate mosses, lichens, or other light- or dark-colored spots at the base of tree trunks. Other larvae or adult insects use similar collective camouflage strategies.

Each habitat offers a variety of forms, textures, and colors that insects may imitate to camouflage themselves during one stage or another of their development. Some imitate a plant's thorns, others mimic its seeds or leaves; in fact, all forms of vegetation come into play. However, matching by color and form is not restricted to the plant kingdom. Rocks and soils are used as camouflage models by several species of locusts that display bright colors while flying, only to disappear as they land on the ground.

Bird droppings, which are imitated by several caterpillars and a few other insects, make a particularly interesting model. Indeed, no better model for insects could be found since most insects' predators are birds, and once their droppings have been deposited on leaves or branches, birds pay no further heed to them. As a result, insects that imitate bird droppings could not find a better form of life insurance. The mysterious origins of cryptic coloration are even more puzzling in this case since bird droppings are an incidental feature and, unlike leaves, bark, flowers, thorns, and twigs, are not an integral part of the biotope.

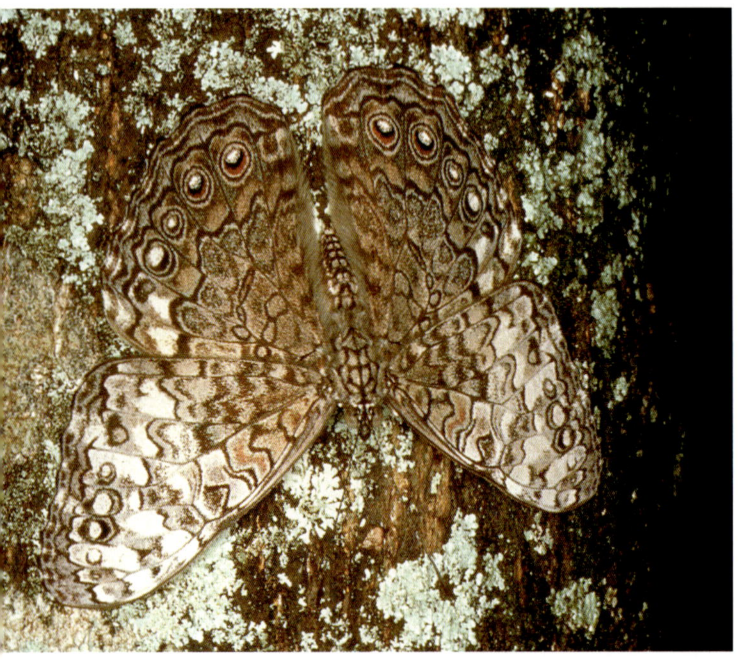

Cryptic coloration is always accompanied by behavior designed to make this camouflage as effective as possible. Insects whose appearance provides camouflage are thus "programmed" to behave in an appropriate way. Those that resemble details of their environment move or fly away only as a last resort.

Above: The gray cracker butterfly *(Hamadryas februa)*, a common nymphalid in the Dominican Republic, always lands upside-down. (See also *Thysania zenobia* and *Thysania agrippina* on pages 100 and 105.)

Camouflage defenses presume that attacks will not take place: either the camouflage is successful and everything is fine, or the game is up. However, the victim of an attack may well have another ace in the hole. *Automeris* moths, for example, surprise their assailants by displaying a flash of color; they then fly away, leaving the startled assailant behind. Some insects also play dead by dropping to the ground, where they usually disappear amid plants and leaf litter. Even if the insect is still visible, the predator may not recognize its motionless form and may lose its attack reflex; this phenomenon occurs among many hunting animals.

Dissimulation also takes the form of disruptive coloration, which relies on jarring colors and patterns that disrupt an insect's overall form, thus rendering it unrecognizable in most contexts. Unless they position themselves on an appropriate background, however, garishly colored insects often risk drawing attention to themselves. This strategy is thus distributed more sparingly than other types among insects and other organisms.

Some strategies, seemingly designed to obscure the image a predator has of a type of prey, anticipate or even invite an attack in the hopes that any eventual

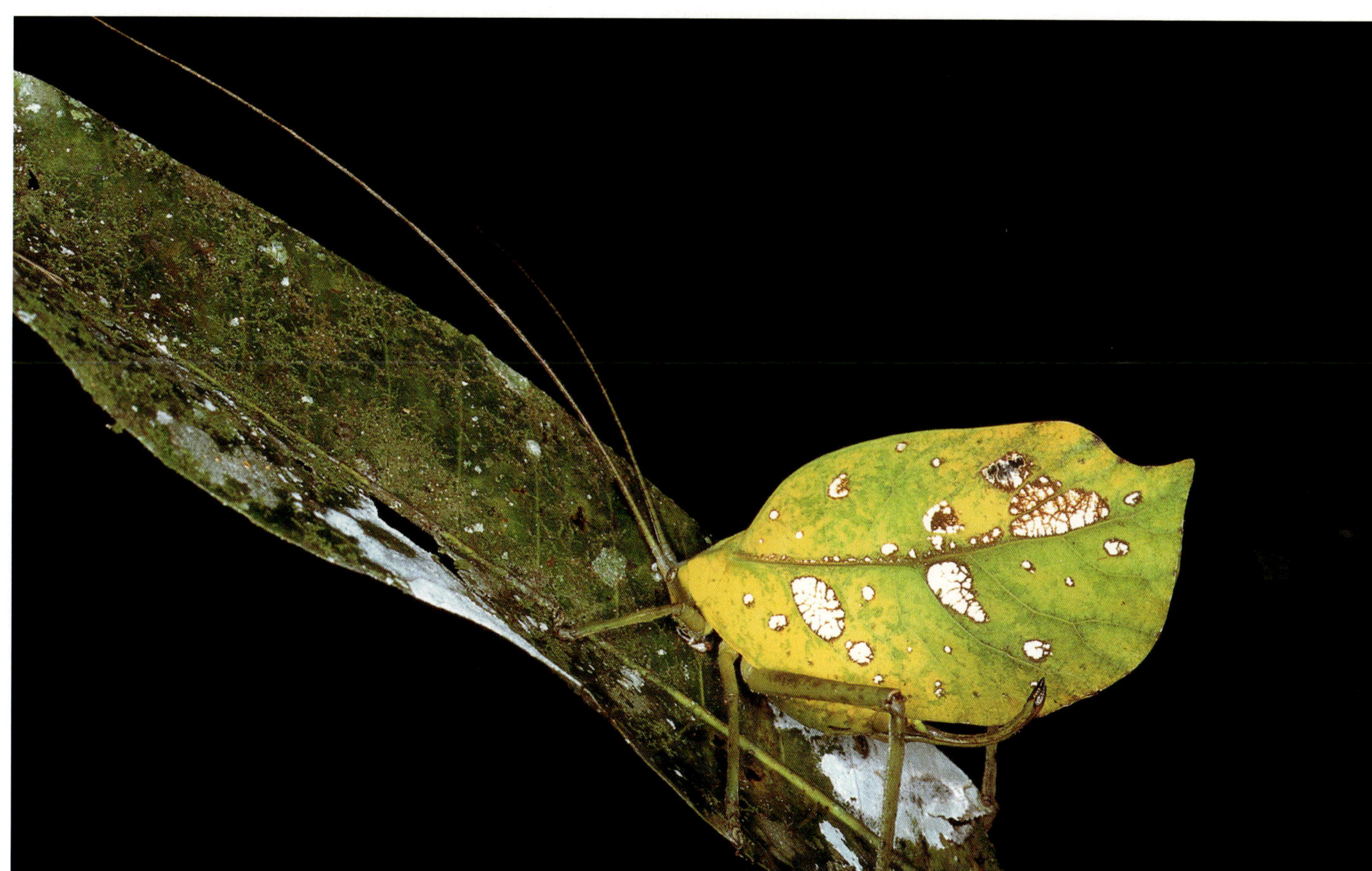

It is one thing for a grasshopper to emulate a leaf. However, when the mimicry includes spots that look like the mold that forms where insects have nibbled on leaves, nature's creative power seems somewhat disquieting.

Above: A grasshopper (family Tettigoniidae) from the Amazonian region of Peru.

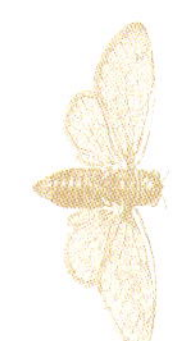

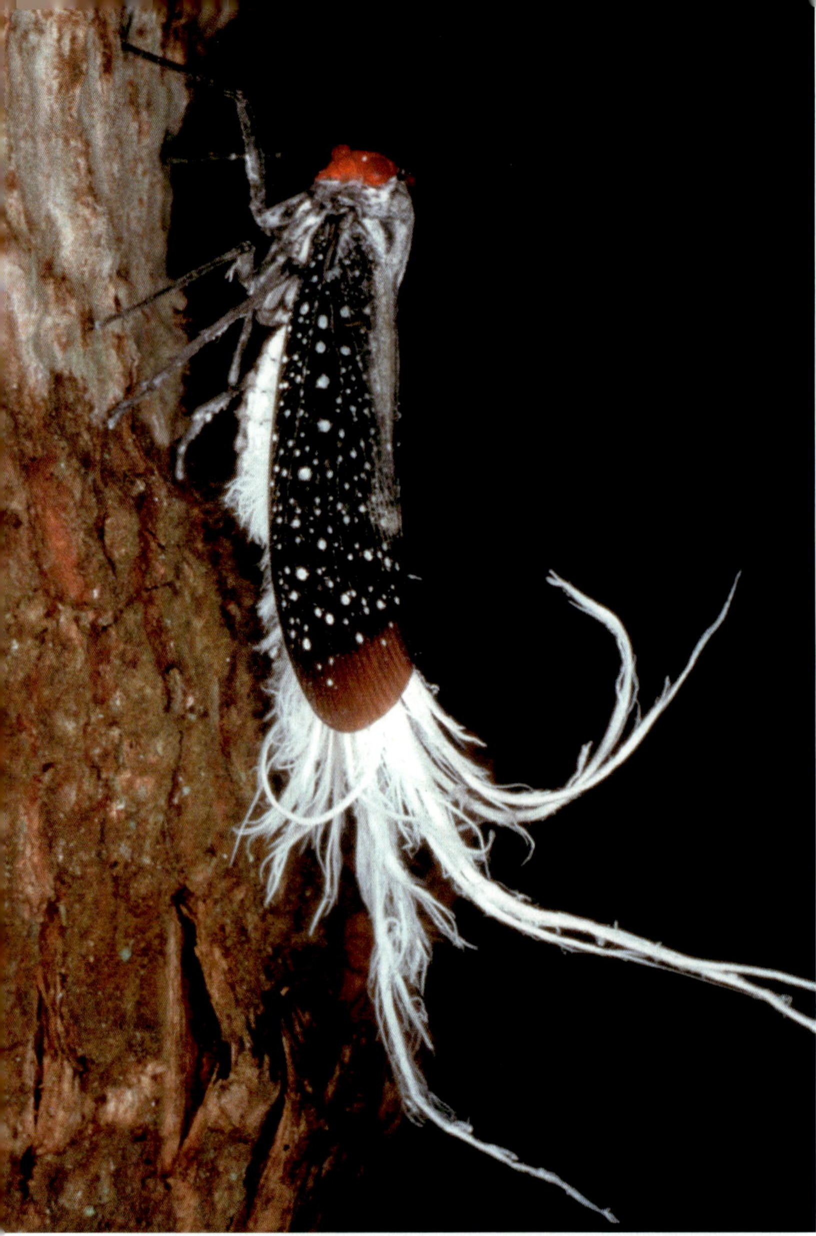

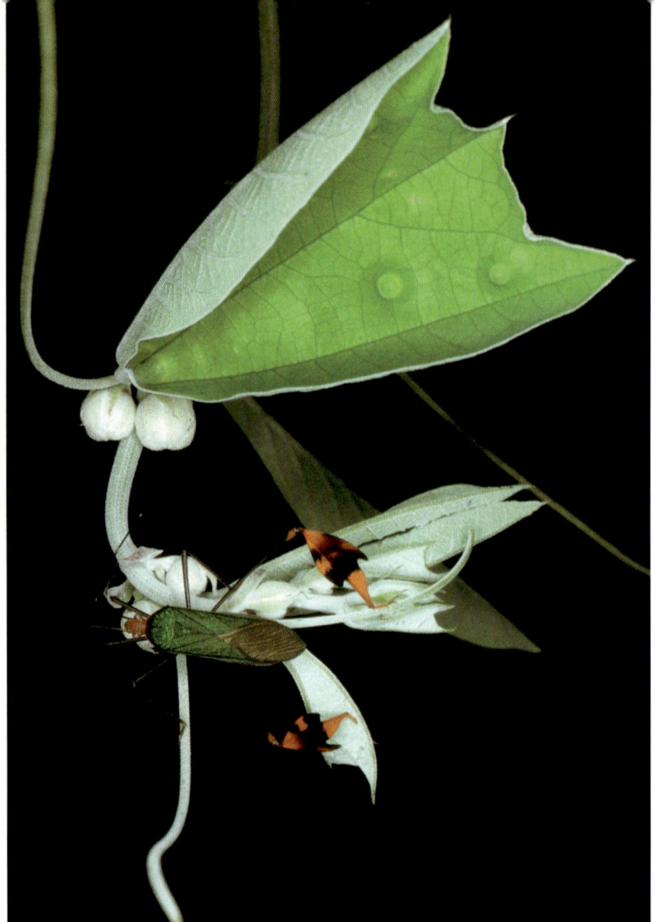

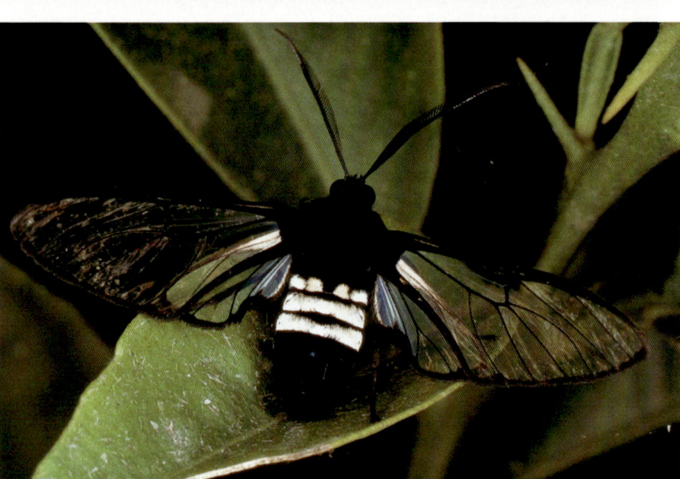

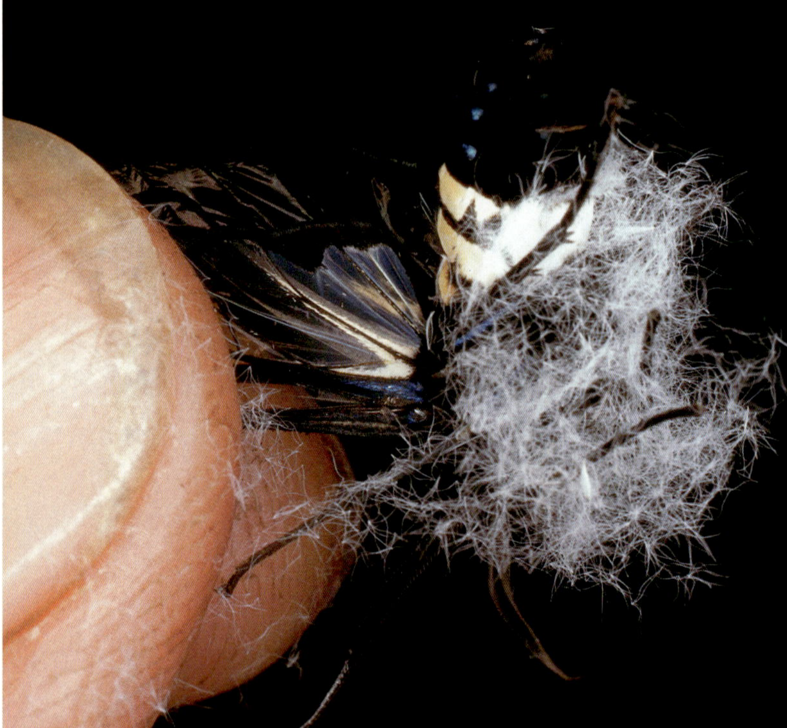

Otherwise defenseless insects exhibit as many camouflage patterns and startle effects as there are species—strategies that allow them to remain invisible or to confuse predators.

Top left: This small homopteran, a wax-tailed hopper *(Lystra lanata),* manufactures "false" ornaments, the only things that will find their way into a bird's mouth.

Top right: Similarly, the *Anisoscelis foliacea* bug, with its foliated legs, attracts the attention of predators that focus on these garish elements, which are sacrificed to protect the rest of the body.

Bottom: When captured, this small ctenuchid moth emits a compound that instantly forms a stringy silk-like substance upon contact with the air. These insects are found in tropical America.

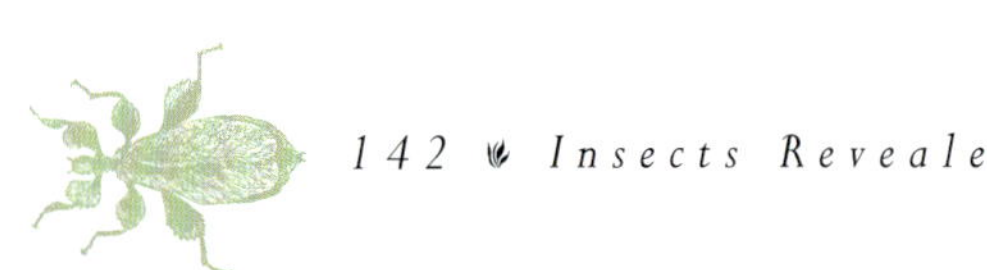

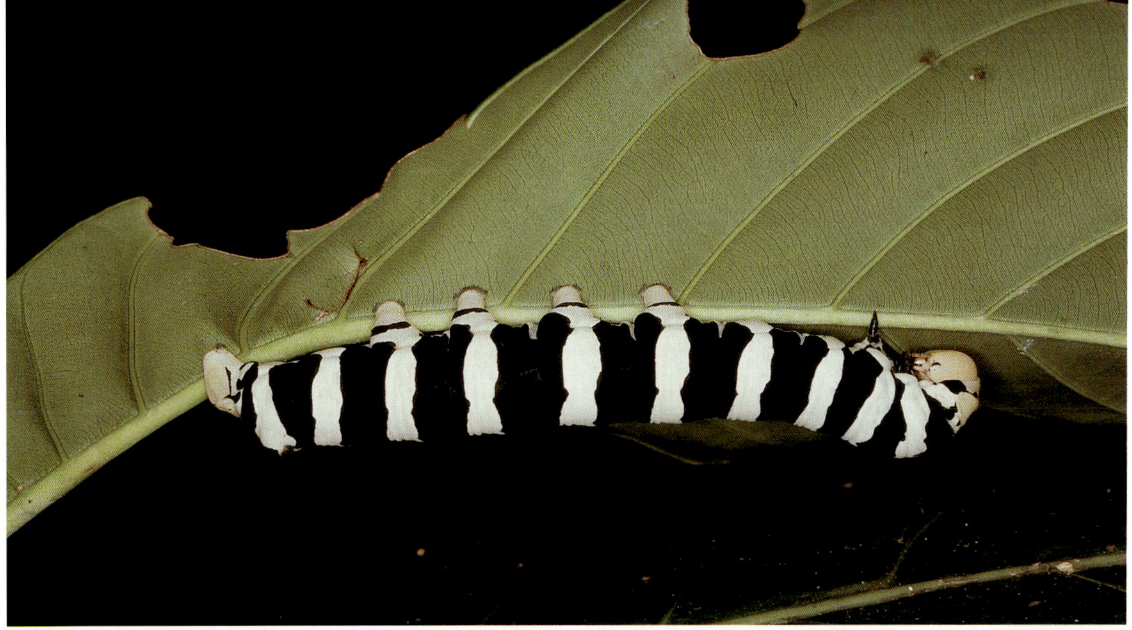

Defense strategies based on trompe-l'oeil are much more diverse and ingenious than are chemical or mechanical defenses. Trompe-l'oeil includes disruptive coloration, which often works in tandem with passive chemical defenses that rely on premonitory coloration. The latter is the only form of protection available to organisms that harbor toxic substances.

Above: You have to look twice before detecting this hawk moth caterpillar, which resembles a series of small white flowers.

injuries will be sustainable. Such tactics serve to deflect or confuse predators' attacks, redirecting their attention to more tempting ornaments on the periphery of the insect's body or to a false head marking on its posterior wings—nonessential body parts that the insect can lose during an attack without perishing. It may even survive a second attack if additional expendable body parts can be sacrificed. By means of this tiny margin of survival, a species may successfully maintain an equilibrium within an ecosystem.

Other insects avoid paying with their own body parts by relying on disguises and false features that serve much the same purpose. Certain larvae cover themselves with bits of wood to blend in with their surroundings; others manufacture structures from their own excrement or skin sheddings, which are retained for this purpose after each molt and build up on the insect's caudal (posterior) spines. What distinguishes such defenses from cryptic coloration is that the insect actually makes its own camouflage. In this crafty category, no insects are as successful as the Fulgoridae (planthoppers) of tropical America, which excrete highly brittle plume-shaped crystals of white wax that grow as long as the insect's body. This comet's tail apparently serves as a decoy; when a bird pecks at the artificial body part, the insect promptly escapes.

These preventive defense mechanisms all rely on some form of appearance-related trickery or camouflage, thanks to which well-disguised insects remain invisible to their predators and do not have to defend themselves actively.

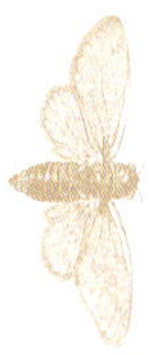

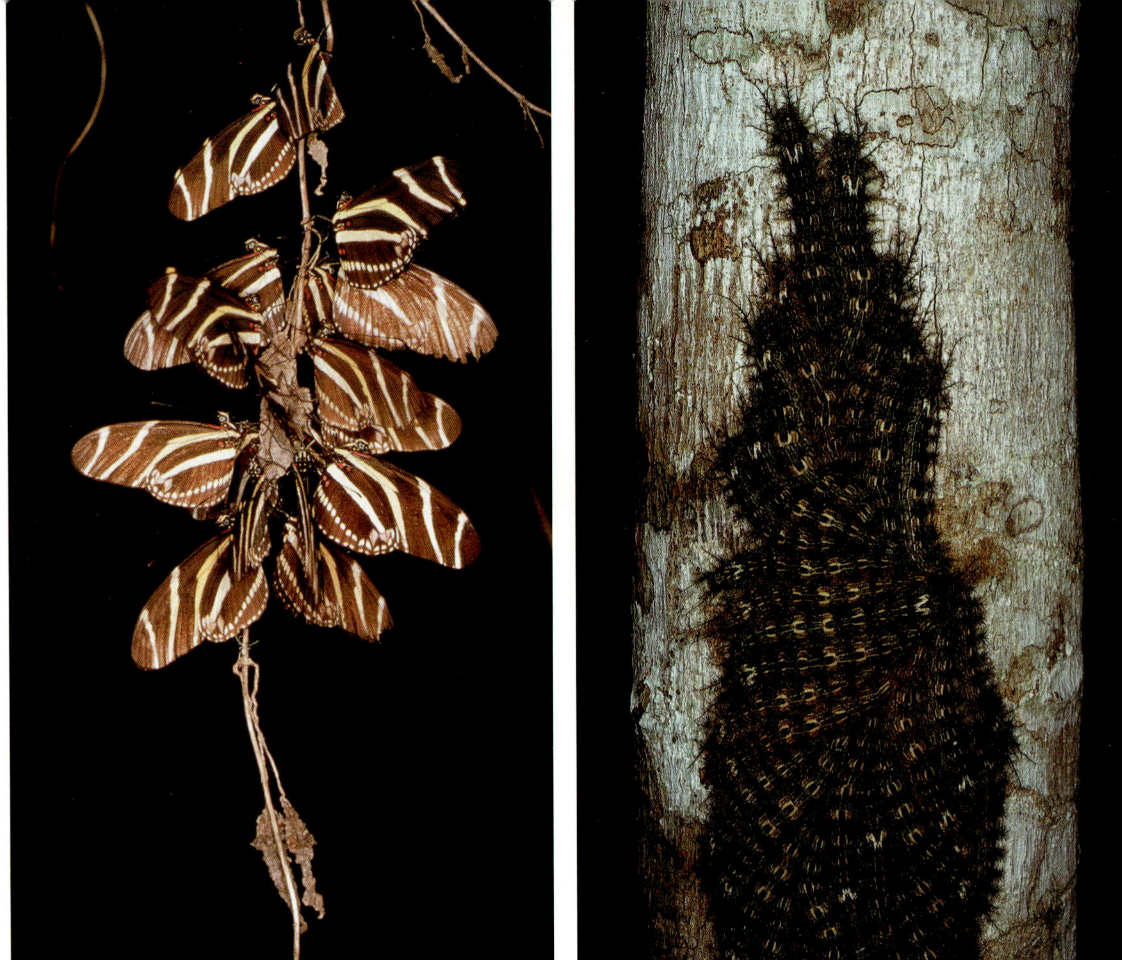

Most defense strategies are used only by individuals; other techniques, however, require group collaboration.

Left: The heliconian butterflies are toxic and use premonitory coloration, which is only effective by day. At night, these butterflies cluster together and warn off predators by their odor, which is strengthened by their large numbers. Shown here: zebra longwing butterflies (*Heliconius charitonius)* from the Dominican Republic.

Right: Tightly packed together, these caterpillars look like moss and are bypassed by predators.

ACTIVE AND PASSIVE CHEMICAL DEFENSES

In addition to such wily hoaxes, there are other passive means of defense that engage a predator more fully because they have an educational function. This is the case for premonitory or warning coloring, which warns predators that an attractive insect is in fact inedible. The insect in question may be foul-tasting, foul-smelling, or impregnated with some lethal poison. On occasion, individual members of these toxic species are sacrificed as part of a young predator's education; this is the price that must be paid for the safety of others.

Many of us remember the acrid smell of ladybugs or the horrible taste of raspberries that carry the residue of a stinkbug (family Pantatomidae). Insects' predators have even better memories of nature's ubiquitous warning signs, and learn at a young age to read and recognize them. However, many insects add a second defensive weapon to their arsenals by clustering in tightly packed groups, thereby reinforcing the olfactory and visual signals they emit.

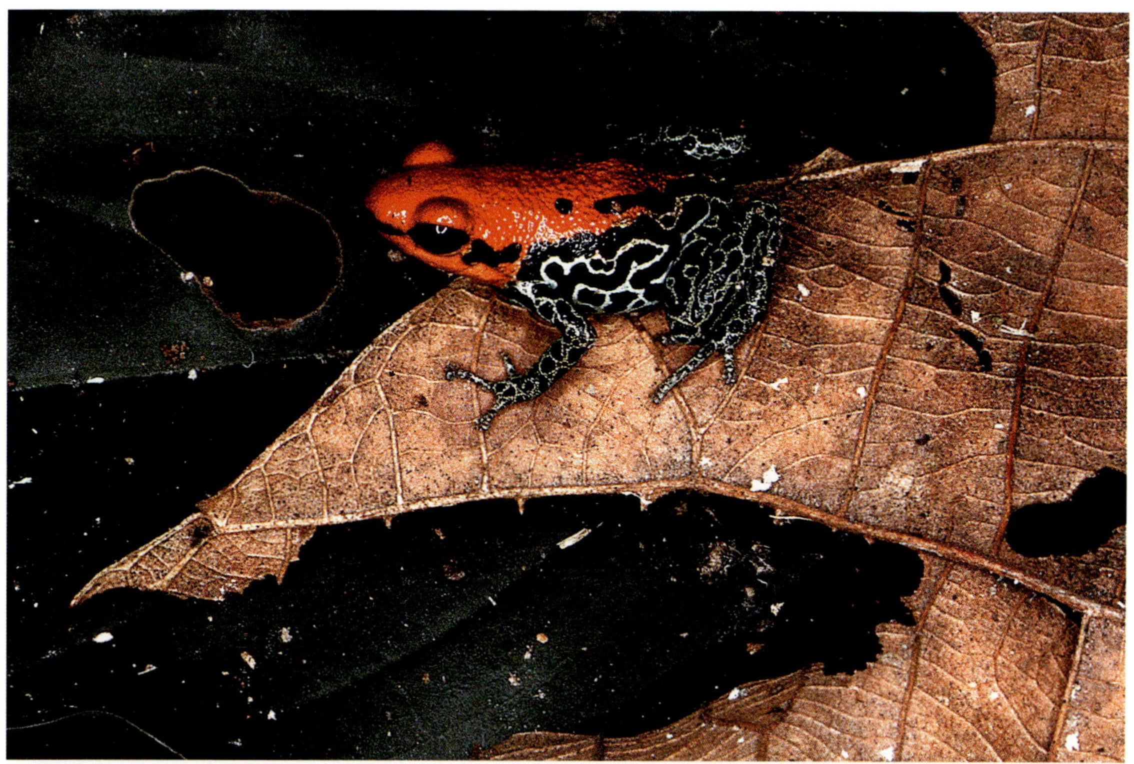

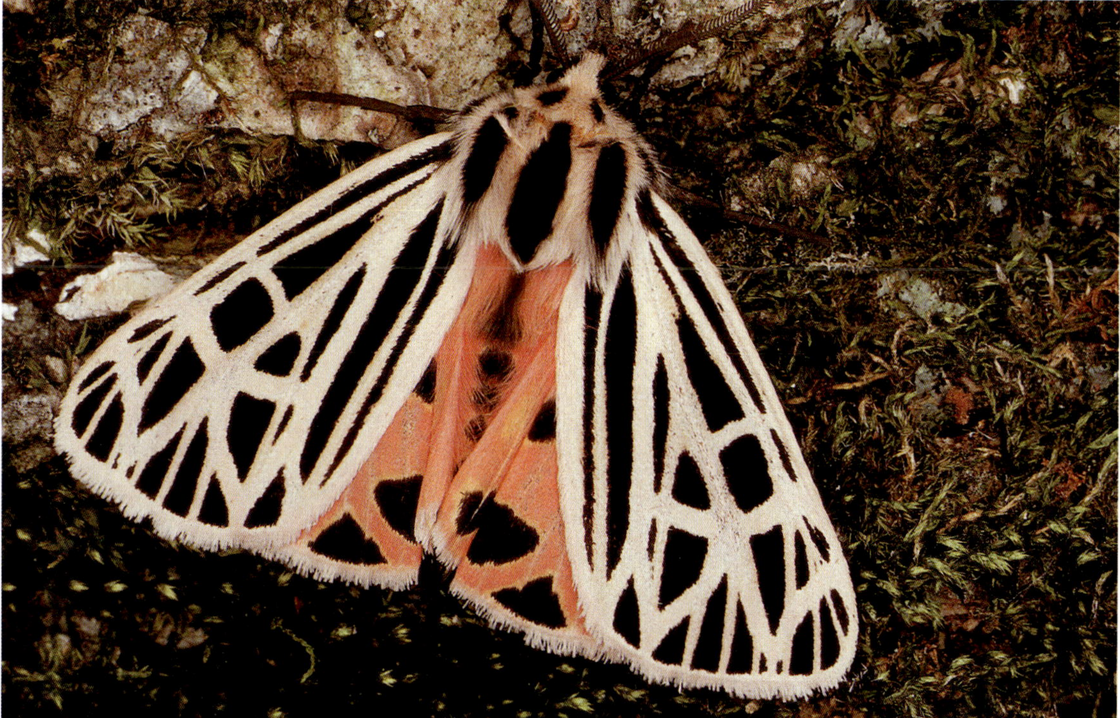

Premonitory coloration works only if the predators understand the warning signs, which are clear to all but young and innocent predators.

Top: A poison dart frog from the Amazonian region of Peru.

Bottom: This moth, *Grammia virgo* (family Arctiidae), is indigenous to northeastern North America.

The poison dart frogs and the *Grammia* are both toxic, but they are unable to transmit their toxins to a predator unless it bites or swallows them. Although the individual is sacrificed, other (often related) individuals benefit since the predator that survives the encounter learns not to touch overly garish specimens in the future.

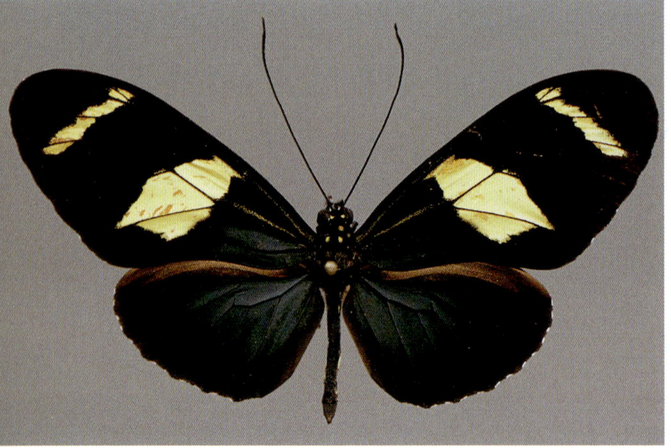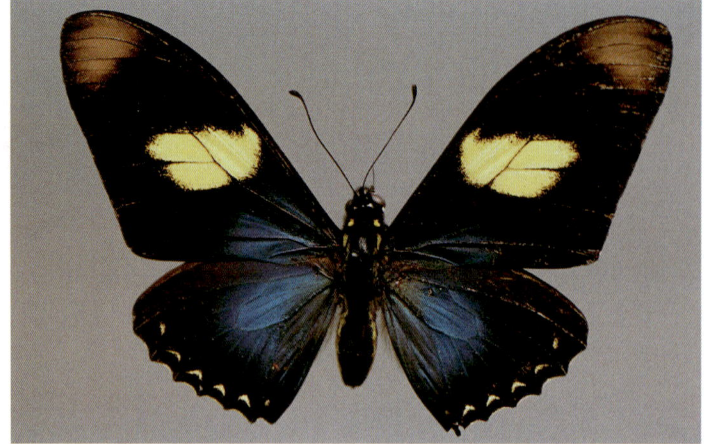

Insects using chemical or mechanical defenses also exhibit the mysterious phenomenon of Batesian mimicry, which was discovered by the British naturalist Henry Walter Bates during his exploration of the Amazon basin in the mid-nineteenth century. This form of mimicry features a model that is harmful, and a mimic that, although harmless, is protected by its similar appearance.
Left: A model: the toxic *Heliconius wallacei* (family Heliconiidae).
Right: A mimic: the nontoxic *Eurytides pausanias,* a papilionid butterfly.

Insects and other animals that contain toxic substances, whether self-produced or absorbed with no ill effects, are normally passive in the face of danger. Seemingly confident, they do not shy away from predators, which should recognize them by their premonitory colors. For example, skunks neither fight nor flee, secure in the knowledge that the only weapon at their disposal is also the most effective one. Some insects that utilize chemical defenses have the means to transmit venom or other irritants to their enemies. These include the stingers used by wasps, bees, and ants, together with short-range mechanisms that emit formic acid or viscous substances, as in the case of carabid beetles and termites.

Several caterpillars have erectile stingers that contain highly toxic venoms; such species are best left undisturbed. Certain "bombardier" carabids defend themselves by spraying chemicals into a predator's face or mouth. The crackling sound of the emission startles the predator, and the burning sensation then puts an end to the attack.

This discussion in no way exhausts the arsenal of means to transmit poisons or of all the poisons, venoms, acids, and enzymes at work. Insects use these substances not only for self-defense but also to paralyze, anesthetize, or digest their prey whole within its own skin, before extracting the liquified nutrients.

Chemical warfare as waged by insects is thus both defensive and offensive.

Defense strategies

PASSIVE DEFENSES

A. Cryptic coloration relies on camouflage or dissimulation; it is sometimes accompanied by playing dead or by the use of startle effects.

B. In cases of disruptive coloration, the overall form is fragmented by means of brightly colored patterns that render the individual unrecognizable at first glance.

C. False identification of vital parts serves to confuse predators by means of decoys or deceptive appendages so that they focus their attacks on the periphery of the body.

D. In premonitory coloration, bright colors warn of chemical defense capabilities. Both passive and active forms of this defense give rise to mimicry.

ACTIVE DEFENSES

A. Bluffing behavior may involve brandishing false weapons or feigning hostile behavior.

B. Authentic chemical or mechanical weapons include noxious substances; stingers or spines on the body or legs; and, most notably, powerful mandibles. In cases of mimicry, "armed" species are imitated by other completely harmless species.

Note: If any of these defenses prove unsuccessful, an insect may flee if it still can. In fact, fleeing is the only means of defense available to many insects.

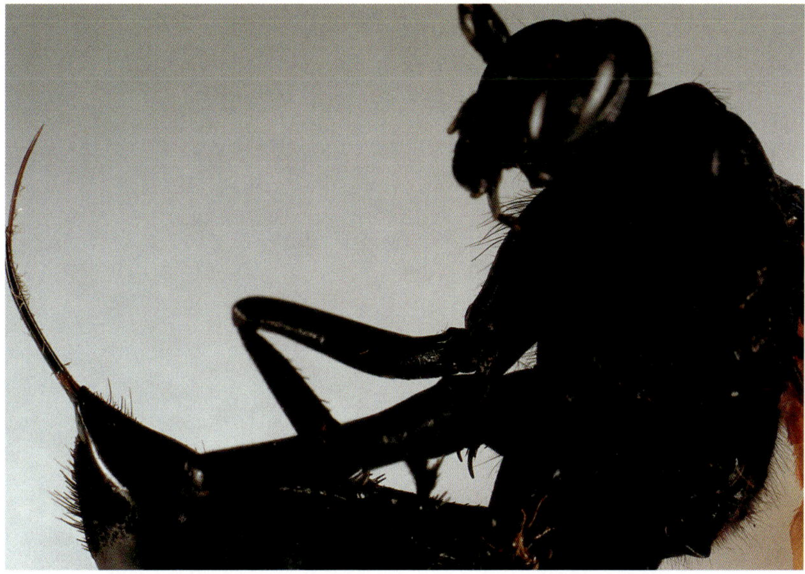

Chemical defenses
Being full of a toxic or repulsive substance can assure active and direct protection only if an organism has a mechanism for transmitting the substance to an enemy by injection, spraying, or some other means.
The stinger of this wasp, *Pepsis* sp. (called "mata catata," or spider-killer, in the Dominican Republic) could be used as either an offensive or a defensive weapon—although I do not know what might prey on it.

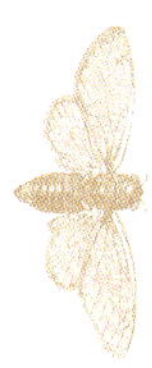

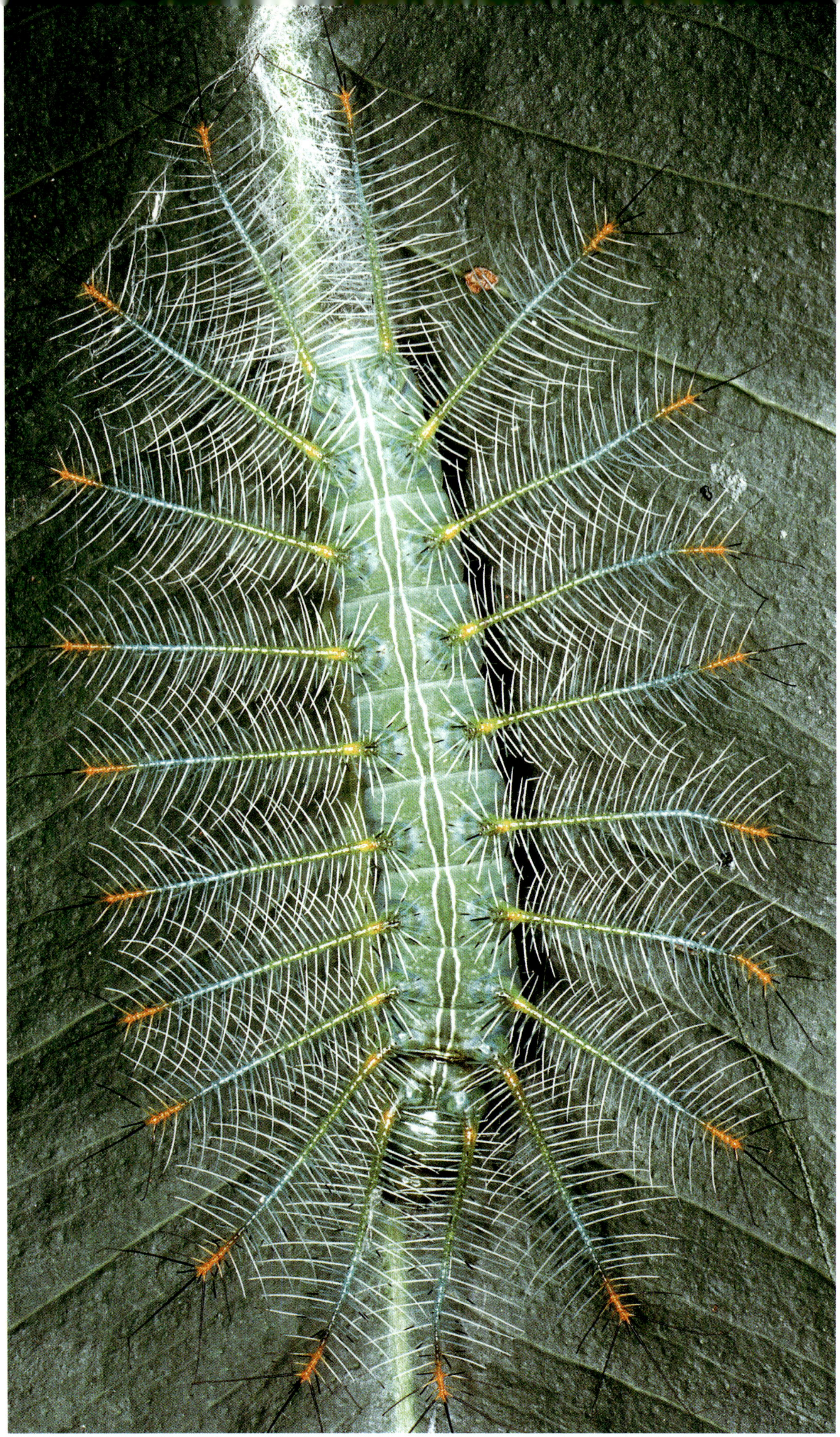

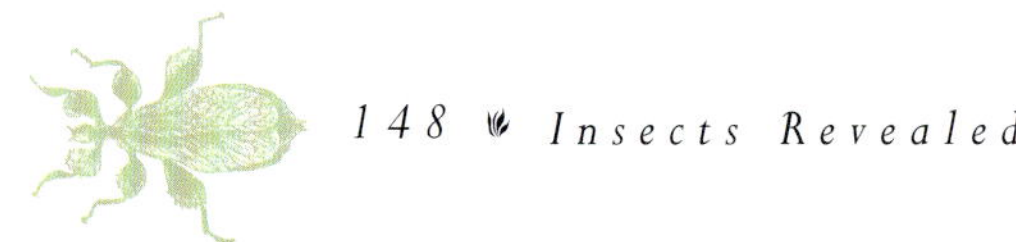

This archduke caterpillar *(Lexias dirthea)* from Southeast Asia is secure beneath its crown of stingers, which resemble hypodermic syringes that are best left untouched.

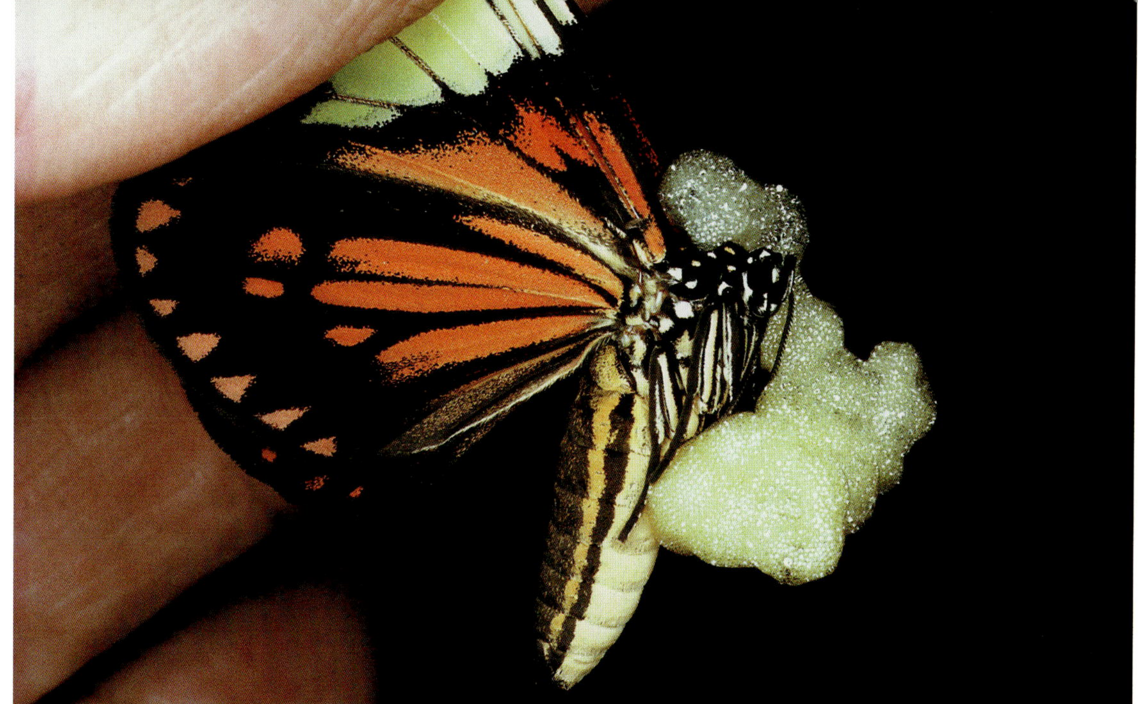

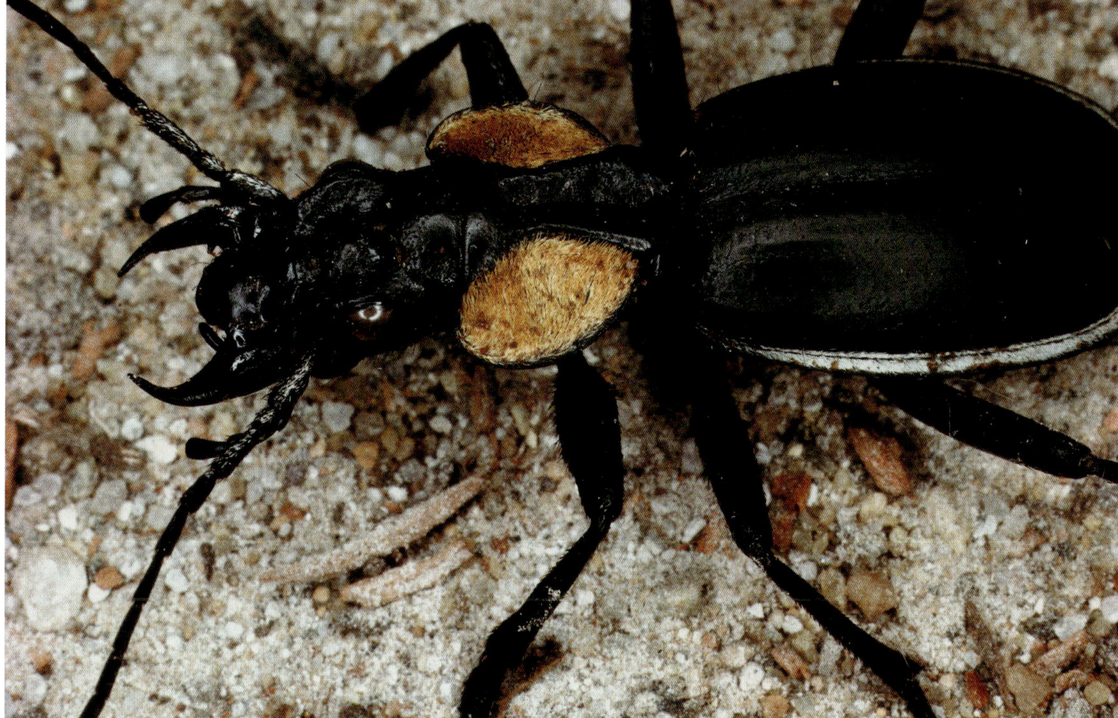

Additional chemical defenses

Top: All tiger moths (family Arctiidae) are protected by an acrid smell that renders them inedible. This chemical defense is signaled by premonitory colors. Some species also exhibit a second means of defense: if touched, the moths excrete a foamy repellent substance that causes predators to spit them out. Shown here: *Chetone phyleis,* an arctiid from tropical America.

Bottom: This ground beetle, *Anthia thoracica* (classified by some taxonomists as a carabid and by others as a harpalid) is indigenous to the semi-arid regions of Africa. Its mandibles are used to capture prey or for self-defense. If attacked, it uses a very effective chemical defense, spraying formic acid for a short distance around itself.

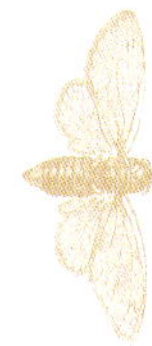

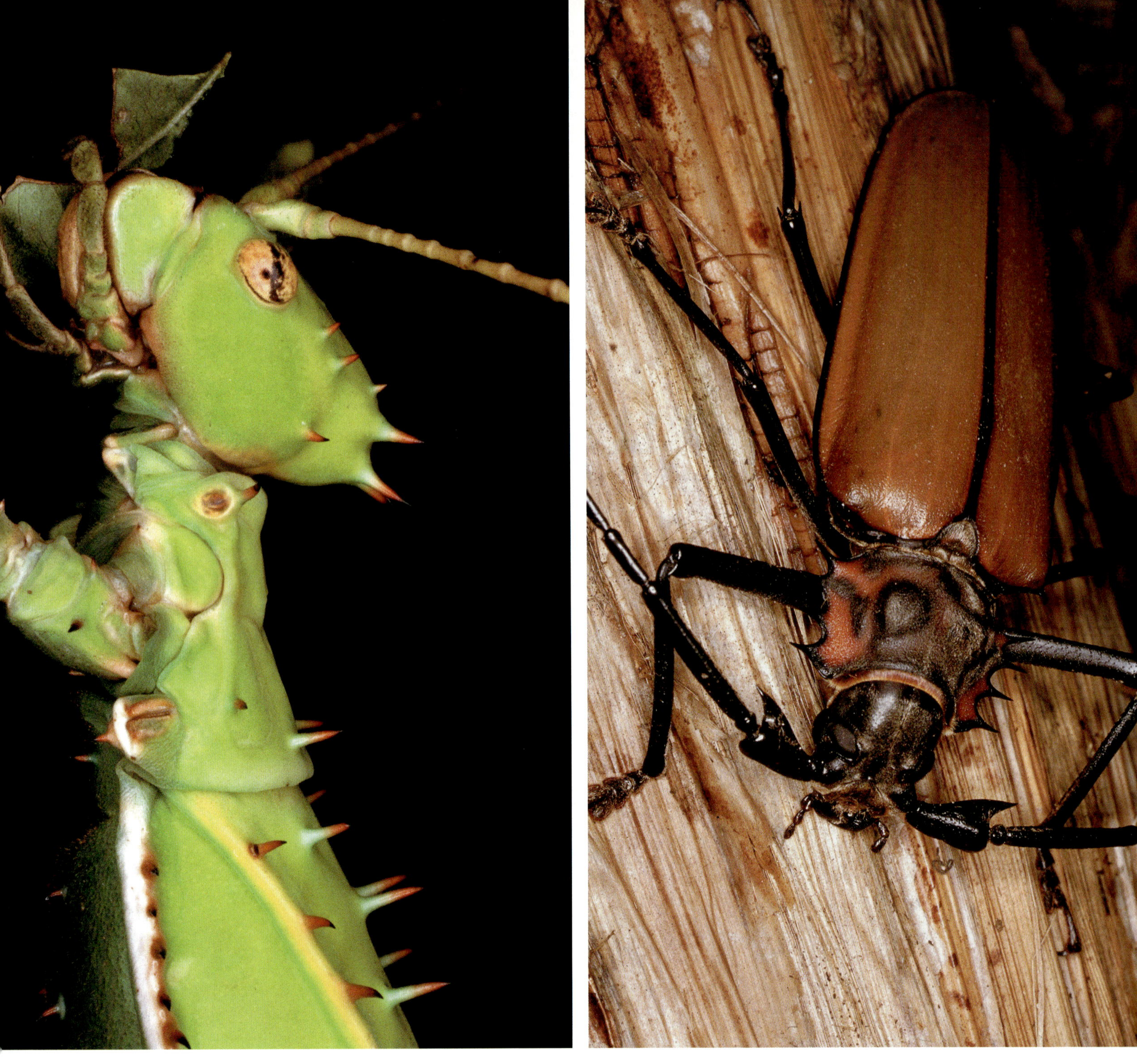

Mechanical defenses

Some insects possess tools, such as mandibles or spines, that enable them to slash, pierce, or otherwise wound predators.

Left: The jungle nymph stick insect *(Heteropteryx dilatata)*. This phasmid is covered with extremely sharp spines capable of inflicting pain on any creature careless enough to touch it.

Right: The prionine beetles do not disturb their neighbors in any way. When attacked, however, their mandibles become terrible weapons. *Callipogon armillatus* (up to 11 cm. [4.5 in.] in length) is the second-largest prionine after *Titanus giganteus*. Both of these giants are found in tropical America.

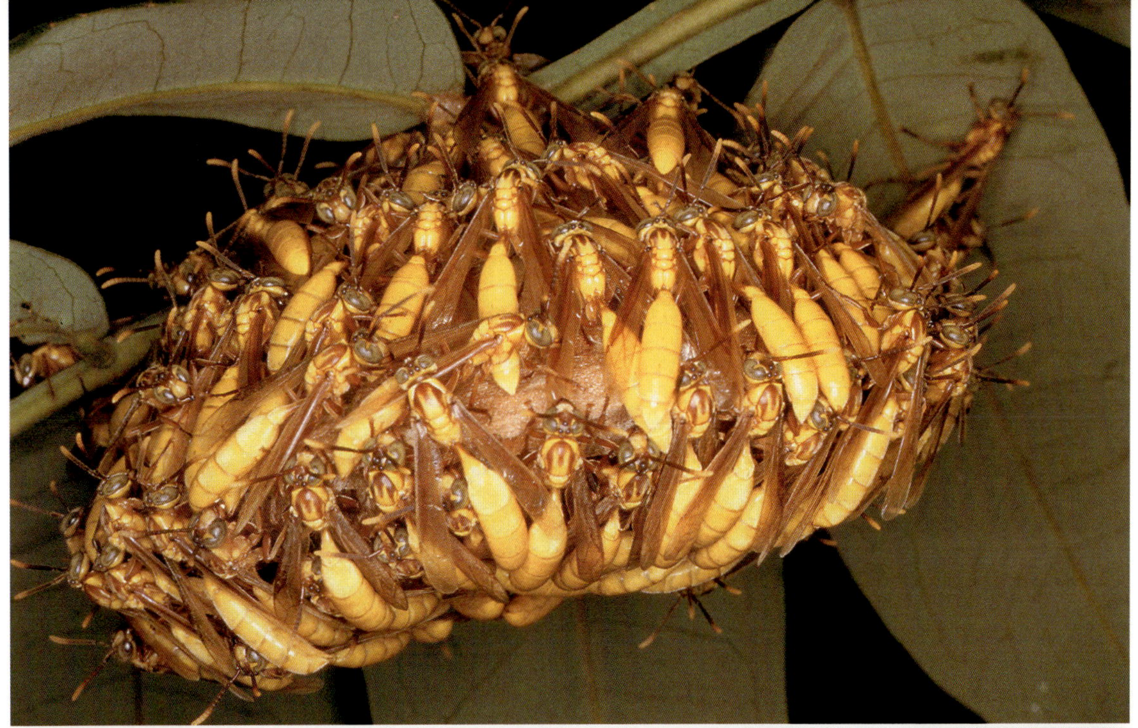

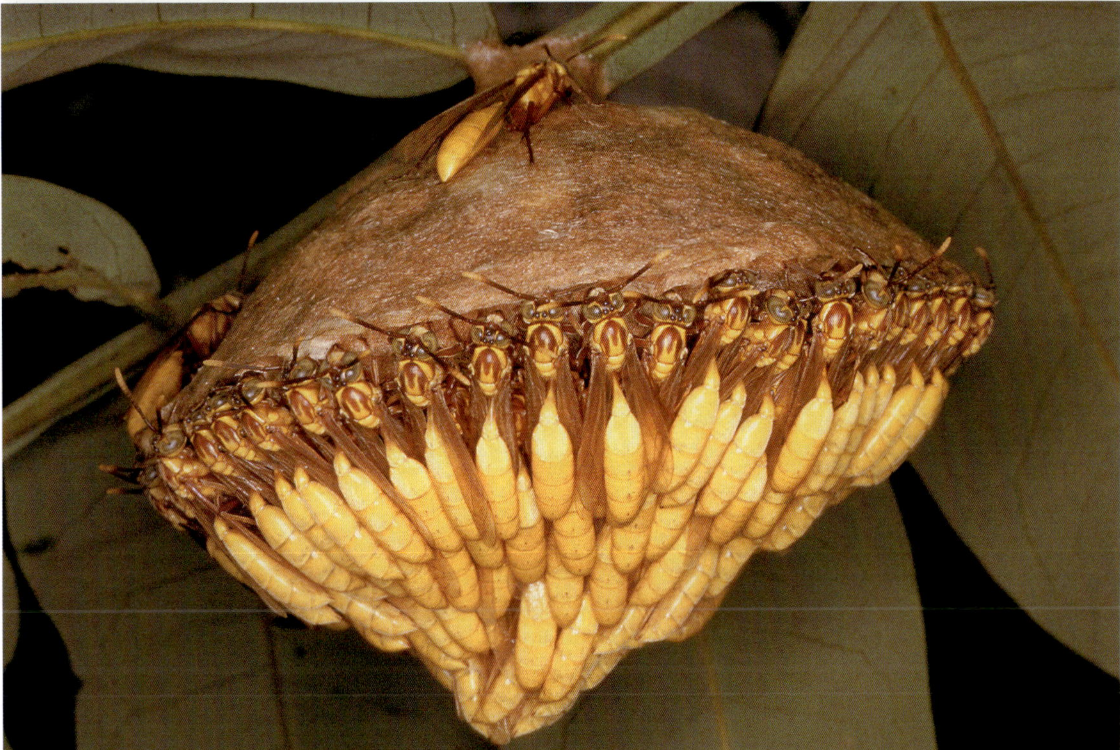

I realize that throughout this book I have paid more attention to the solitary insects than to the social insects, such as bees, wasps, ants, and termites. Perhaps this reveals something about my own temperament. Nevertheless, I would like to pay tribute to the admirable social instincts that these creatures have developed over the ages and that have led them to live in hierarchical communities ruled by queens. Within such communities, individuals per se do not exist; they are merely cogs in a superorganism.

These two photographs illustrate this aspect of the social insects. They were taken in Brazil early one morning when these nocturnal wasps *(Apoica pallens)* were returning to their nest to sleep during the day. I took the second photo some twenty minutes after the first, when I realized that a different type of organization had taken shape within the group. An order had been given and each individual took its place, forming a living shield along the underside of the nest, which was exposed to the air and to diurnal predators.

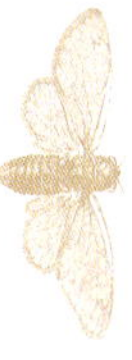

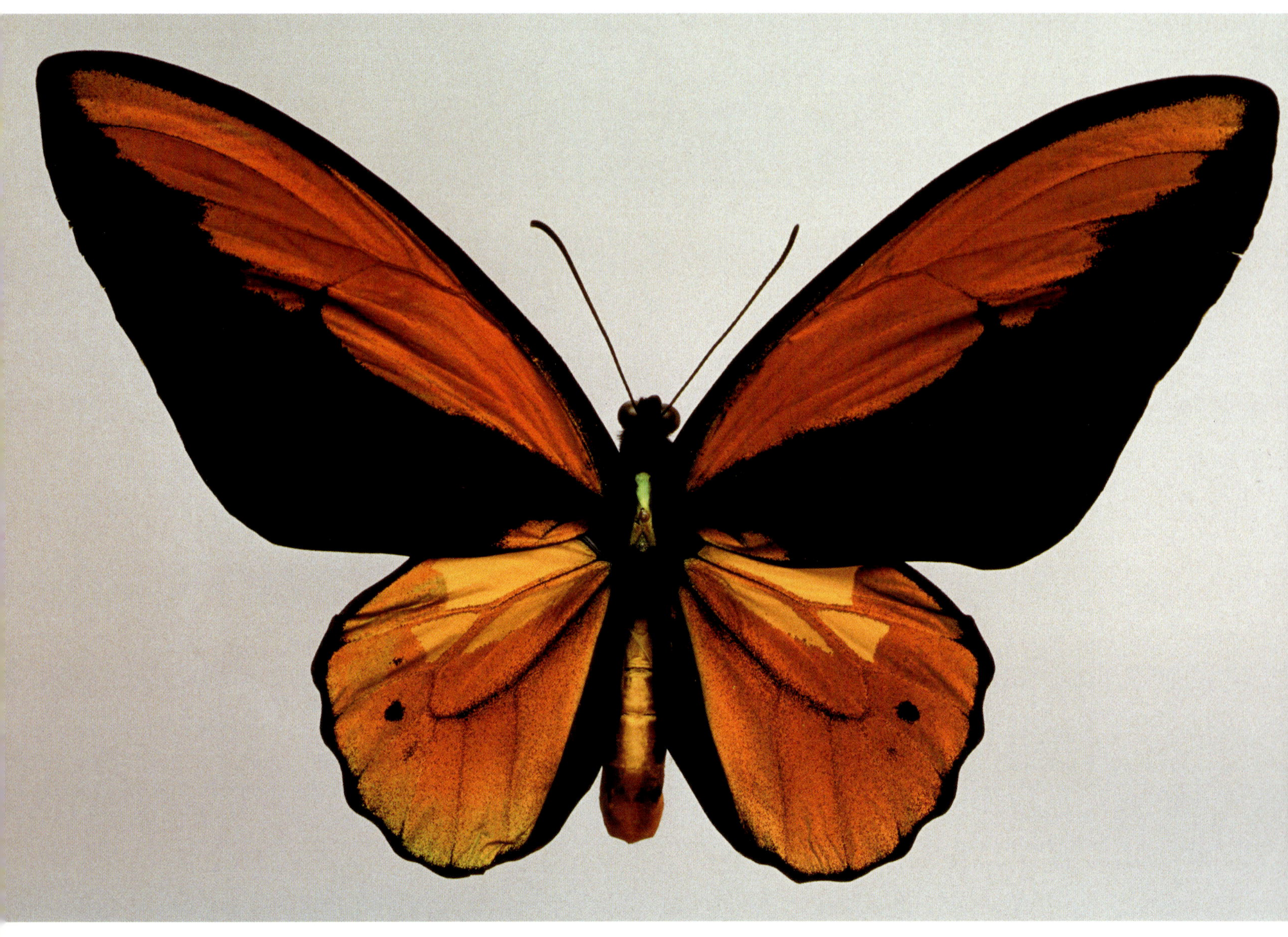

Finding a gold birdwing butterfly *(Ornithoptera croesus)* was an unforgettable moment in the career of the great British naturalist Alfred Russel Wallace. Novice collectors may also experience such moments of glory even if their specimens are more modest and are already well known to science. All that matters is that the specimen be as yet unknown to the collector.

13

the strange need
to collect

After watching one's neighbors or relatives go to incredible lengths to chase insects, only to pin them and store them in boxes according to some scientific convention, one may indeed wonder, why bother?

Ernst Jünger, a German writer and prominent insect collector, describes the hidden attraction of these creatures that many would like to see exterminated: "They are the miniature window through which [the collector] perceives the splendor of the universe."[8]

The desire to collect insects often appears during childhood, long before we can comprehend the passion that has seized hold of us; if this passion lasts, it will end up influencing several different levels of our personality.

Collecting stems from our animal instinct to intercept and capture things that run, fly, and jump across the space that we inhabit, especially when they fall within our immediate grasp. Even though this reflex usually fades and becomes secondary to more intellectual needs, it may still prove helpful on occasion, and the primitive pleasure that we experience when we capture something remains with us always. All collectors have cherished memories of incredible feats of skill that netted them various specimens. Of course, collectors become less bold and less athletic with age. As for myself, I no longer chase after butterflies or other insects that can fly faster than my

---

8. *Chasses subtiles* **(Paris: Christian Bourgeois, 1969).**

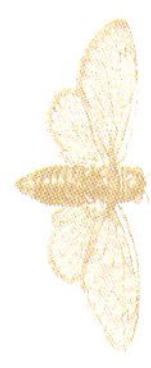

legs can run. To make up for this, I tend to hunt and gather nocturnal insects that I attract with my light trap. Since nocturnal insects are far more numerous than their diurnal counterparts, I lose nothing on the exchange; they alight on the white cloth of the trap and can then be plucked like flowers. Some may deem this a rather mild-mannered form of hunting, especially if they have never experienced the thrill of it. Even if it lacks the edge of full-fledged competition, nocturnal insect hunting is exciting nonetheless since it abounds in mystery and surprise.

With its roots in the instinct to capture things, how does the passion for collecting turn into an intellectual endeavor?

Even though human instincts play an important role in the development of a need to collect, they are not the decisive factor. Strictly speaking, people do not always "capture" the items in their collections; collectors "capture" stamps and rocks only in the figurative sense of the word, as do insect collectors who put together collections thanks to catalogs from dealers in France, Papua New Guinea, Malaysia, or California. For these collectors, a new catalog is like a new hunting ground, as worthy of discovery and exploration as a new biotope. Everything is relative, and perhaps these collectors qualify as hunters in their own right as well.

What most surely propels a person to begin collecting is a subtle process that one does not become aware of or understand until it is too late to turn back.

Here is a typical scenario: you find something that amuses or intrigues you, and then put it on some shelf or in a drawer. However, you do not know what to do with it until you can determine its relationship to a broader category or family; only now does it acquire some meaning. If you come across a second amusing or intriguing object—similar to but in some ways different from the first—you may well feel the desire to examine these similarities and differences. You are not a true collector, not yet, but you are well on your way.

The two objects in your possession may have fired your curiosity enough that you are fatally driven to find a third. Now you are done for, having succumbed to the lure of the number three. Even if you are no numerologist, the magic powers of this number will soon become evident, and you are now at the beginning of a series that must be pursued at all costs. Because you already have three items in your possession, you have to find even more additions for your "family." Your collection is beginning to form; indeed, a terrible compulsion has taken root.

Little do you know that you are reliving, in another form, the legend of the Flying Dutchman's phantom ship, condemned to travel the stormy seas without respite.

All exaggeration aside, collecting is not a truly heavy burden to bear. The torments of this particular passion are usually transformed into joys that can be found nowhere else. In going from a state of ignorance to one of knowledge, you follow the same path as the one that led from primitive humans first to Aristotle's attempts at classification, then to Linnaeus's successful system of nomenclature, and finally to the latest discoveries of modern researchers.

In his admirable work on butterflies, Paul Smart describes the growing desire to collect in these terms: "Through the centuries, intellectuals had amassed cabinets of curiosities in a more or less random fashion. Bones, fossils, seeds, shells, insects and indeed any curiosities of nature were eagerly hoarded and often earnestly discussed, but with no clearly defined objective. The wealth of material brought back from the far corners of the earth during the seventeenth and eighteenth centuries not only inspired men of science to record and classify these new wonders, but demanded a new and more systematic approach to the study of nature in order to make sense of it all."[9]

Collecting thus involves bringing together interrelated things in order to both determine and assign meaning through comparing and contrasting them. This leads to starting or extending one or more series of objects. In the case of insect collections, the number of possible series is astronomical. How can novice collectors fill in all the gaps in their collections, even if they are only planning to collect specimens from their local area? Fortunately, the innocent collector has no inkling how plaintively these gaps will cry out to be filled. Some missing items will torment collectors for years before they are able to find the species or subspecies that completes a genus.

At the outset of the adventure, a collector's first specimens will prompt him or her to seek out various reference books because each creature must have a name. This research reveals the existence of species of which the collector was previously unaware, but which then become objects of desire. Once under way, this process of give-and-take proves endless. The collector living in temperate regions eventually discovers the world

9. *The International Butterfly Book* (New York: Crowell, 1975), pp. 88–89.

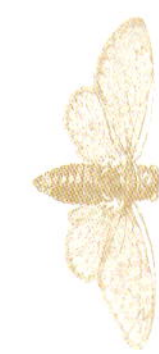

of tropical species, in all its vast diversity. But how does one acquire such fragile and hard-to-find specimens? The collector then learns that the world's insects can be mailordered like other merchandise if a helpful friend can provide the right addresses.

Better still, collectors may manage to break out of their geographical confines in their quest for insects in other parts of the world, species that they would never have imagined flying within reach of their nets: an *Attacus* moth in Malaysia, a *Morpho* butterfly in the Amazon, perhaps even some rare butterfly in Papua New Guinea. Several years ago, I would never have imagined that I would travel to the Amazonian regions of Peru and Ecuador as if I were visiting my second home in the country. I have made so many extraordinary discoveries and had so many unforgettable experiences in these faraway places and elsewhere in the tropics. And it is not just amateurs who feel such raptures. A. S. Meek, the great nineteenth-century professional collector, who discovered numerous insect species in New Guinea, was jubilant when he obtained a male specimen of the species known today as *Ornithoptera chimaera* (a birdwing butterfly), saying, "I was happier than if I'd inherited a fortune." And Meek had not even caught the butterfly himself—he had gotten it from one of his helpers.

In the following oft-quoted passage, Alfred Russel Wallace describes his reactions after having caught an *Ornithoptera croesus* (then unknown to science) on Bacan Island in Indonesia: "This insect's dazzling beauty and brilliance are indescribable, and only a naturalist can understand the intensity of the excitement I felt when I finally captured it. Taking it out of my net and opening its splendid wings, my heart began to beat violently; blood rushed to my head and I felt closer to fainting than if I'd feared certain death. I had a headache for the rest of the day, so great was the excitement produced by what will seem to most people as an inadequate cause."[10]

Novice amateur collectors can experience such moments of glory as well, even though the specimens they capture are more modest and are already well known scientifically; the only prerequisite is that the species be unknown to the collector. Collectors' rewards are commensurate with their own desire: the keener the passion, the greater the reward.

With each passing day, collectors add something new to the "organisms" they are creating or to the structures they are building; these may well become the focal point of their very existence, perhaps their place of refuge. Collections tend to grow as

---

10. *The Malay Archipelago* (London: Macmillan, 1869).

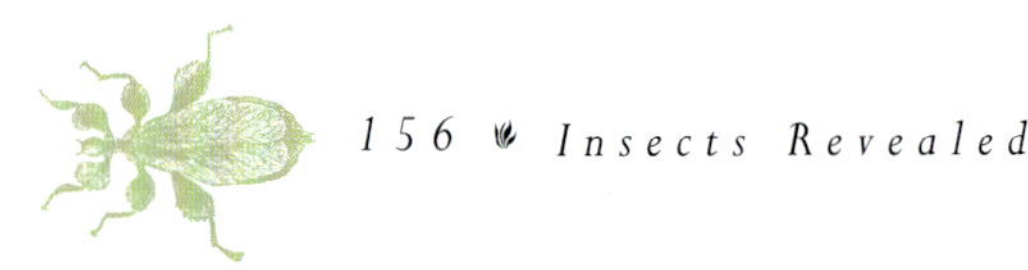

empires do, eventually encircling their creators; collectors come to feel as if they have not lived unless they have added to their work-in-progress every day. Even the least compulsive collectors sometimes feel that if their collections are expanding, then they are alive, making progress, and conquering new territory. Nothing can clip their wings, least of all oblivion, since collectors secretly hope to outlast their collections in much the same way as famous artists outlive their creations. However, very few collectors have seen their names eternally associated with assorted boxes of pinned insects. All the others have vainly cultivated a fetish made out of their avocation. The problem is this: how do we cultivate our noble passions without letting them control and destroy us? Finding balance in the pursuit of this passion is not always easy, and unsuspecting collectors may begin to show signs of mental instability.

Collectors often look forward to the arrival of an ideal visitor so that they may put their personal collections on display. Every new addition to their opus brings the hope that their special guest will be wholeheartedly enthusiastic; such a reaction would make up for all the stress and strain of worldwide expeditions, not to mention their financial cost. Ideally, this utopian visitor would examine every item in the collection with the same blissful attention as the collector, all the way to that tiny specimen tucked deep inside the very last drawer.

But does anyone so receptive ever drop in?

Perhaps we should acknowledge that these guests might actually be figments of collectors' imaginations.

In the space of an hour or two, how could our guest of honor ever understand and appreciate the collector's experiences chasing specimens in the Amazonian forest, exploring rural Chinese grottoes, or scurrying beneath the street lights of Tanah Rata in Malaysia's Cameron Highlands? What of the enormous wasp *(Vespa mandarina),* captured as it perched on a flowering shrub in the park of Japan's Himeji Castle? This particular feat required darting across a lawn on which no sane person would ever set foot. Could the guest ever imagine the tongue-lashing delivered by an elderly gardener, outraged at such sacrilegious behavior? Each specimen has its tale to tell. This hidden dimension is fated to disappear along with the collector; perhaps even sooner, since, with advancing age, a collector's recollections may begin to fade from memory.

It is frustrating to admit that the more a collection expands, the harder it is to display it properly. Aside from the collector, no one has the patience to rummage lovingly

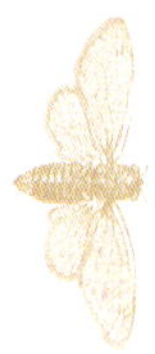

through two or three drawers while the morning coffee is brewing or before going to bed, admiring some brilliant scarabs or long-forgotten Saturniidae. How fortunate they are, those novice collectors who can easily inspect their entire collection because it still fits in one box; those with larger collections are forced to appreciate them detail by detail. Such collectors eventually learn to accept that the ideal visitor is found only in their dreams.

So what purpose do collections serve? This question has tormented me for many years; I now believe that it is the collectors who are the beneficiaries of their collections. From the outset, collectors build and maintain their collections for their own pleasure; this leads them to discover and learn a great deal about insects, about ecology, and even about themselves.

The French poet Paul Valéry describes an imaginary conversation between the souls of Socrates, Phaedra, and the ancient Greek architect Eupalinos, who offers this telling remark: "Through constructing, I believe I have constructed myself."[11] According to Eupalinos, if the collector makes the collection, then the collection also makes the collector. Collections serve as instruments for study, research, and reflection. Indeed, collectors may refer to their collections whenever they need to, and this helps shape their thinking as entomologists. (I know several collectors who turn down any specimens that they did not personally catch; if they did accept them, they would be undermining the notion of authorship and individual creativity that is intrinsic to their collections.)

For the amateur collector, esthetic aspects take on considerable importance as well, and many collections have their roots in the simple desire to admire a handful of beautiful butterflies or shiny beetles. Those who find themselves absorbed by insects' forms and colors risk becoming caught up in a world of fascination that will beckon them even further. Amateur collectors, whose initial concern was merely to satisfy a few whims, soon take pleasure in accepting and subscribing to the requirements of science. Such enlightened collectors discover that they feel the same sense of personal obligation as does a musician interpreting Bach or Stravinsky. Interpreting nature or music calls for the same respect. Taxonomy gives collectors the means to understand the natural order of things and allows them to replicate it through the specimens they collect.

In the end, as Paul Smart maintains, we may be able to make some "sense of it all."

---

11. *Eupalinos ou l'architecte* **(Paris: Gallimard, 1924).**

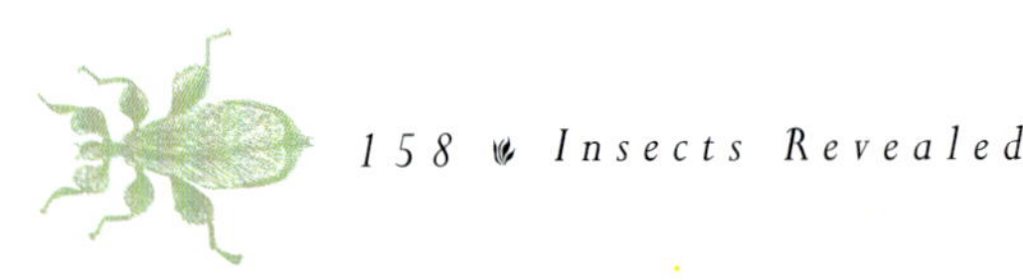

glossary

Apterygots   Primitive wingless insects and hexapods.

Biocenosis   The collection of living creatures that form an ecological community.

Biotope   A small biological region that forms its own distinctive community.

Chrysalis   The pupa of a butterfly; also the pupal case.

Coprophagous   Feeding on dung.

Dimorphism   Morphological differentiation between males and females of a species (sexual dimorphism) or between individuals of a species at different seasons (seasonal dimorphism).

Ecdysis, or molt   The shedding of the exoskeleton in insects and other arthropods.

Ectoparasite   A parasite that lives on the outer surface of its host.

Elytra (sing. elytron)   The horny front wings of beetles and some other insects, which cover and protect the hind wings.

Hemolymph   The circulatory fluid of an insect, consisting of an almost colorless liquid that envelops the internal organs.

Imago   The adult insect.

Integument   The outer protective layer or covering of an insect's body.

Iridescence   A play of rainbow-like colors that shimmer and change as the observer's position changes.

Larva   The immature stage, between egg and pupa, of an insect that undergoes complete metamorphosis.

Nymph   The immature stage of insects, such as dragonflies, that undergo simple metamorphosis. The nymph develops into the adult without a pupal stage.

Ocellus (pl. ocelli)   The simple eye of insects and some other invertebrates, consisting of light-sensitive cells. Also, an eye-like marking on the wings of certain butterflies.

Ommatidia (sing. ommatidium)   The cone-shaped units that make up the compound eyes of insects and some other arthropods.

Paleobotanist   A botanist who specializes in the study of fossil plants.

Parthenogenesis   A type of reproduction occurring in some insects (especially phasmids) in which the egg develops directly into a new individual without fertilization.

Pheromone   A chemical substance emitted externally that affects the behavior or physiology of other individuals of the same species. Pheromones play an important role as sex attractants and are highly developed among the social insects.

Phytophagous   Feeding on plants.

Pterygots   Winged insects, or insects descended from winged ancestors.

Pupa   The third stage of metamorphosis, coming between the larval stage and the adult stage.

Symbiont   An organism living in a state of symbiosis, that is, in a close and mutually dependent association with another species. Some insects harbor symbionts in their digestive tracts which facilitate digestion.

Vestigial   Reduced in size or function during the course of evolution.

Xylophagous   Feeding on wood.

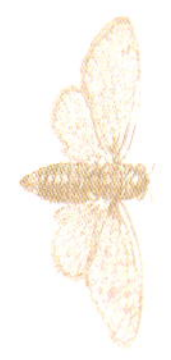

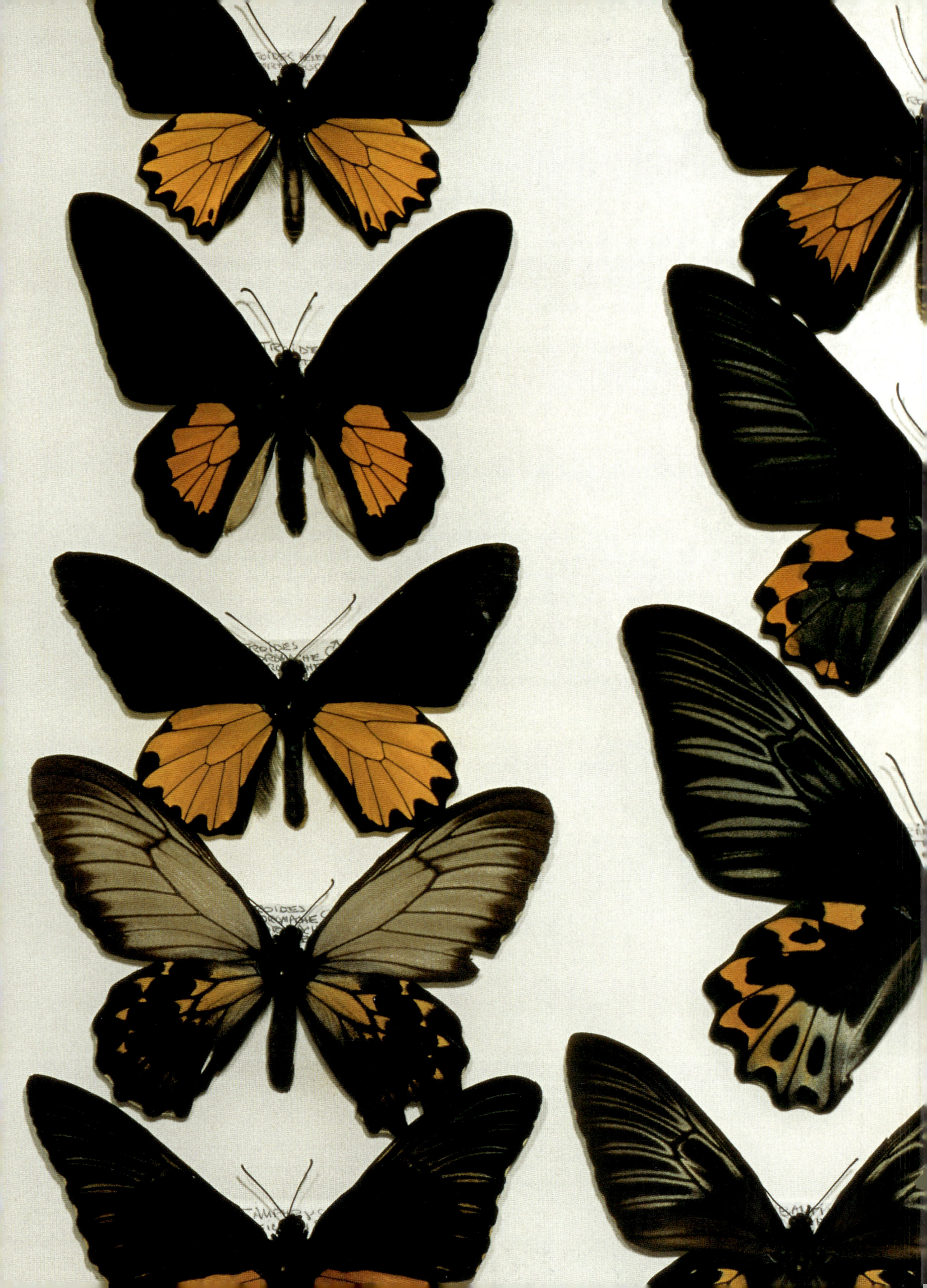